Management Division

Core Programming Topic at the 2011 AIChE Annual Meeting

Minneapolis, Minnesota, USA
16-21 October 2011

ISBN: 978-1-61839-726-3

Printed from e-media with permission by:

Curran Associates, Inc.
57 Morehouse Lane
Red Hook, NY 12571

Some format issues inherent in the e-media version may also appear in this print version.

Printed by Curran Associates, Inc. (2012)

AIChE
3 Park Avenue
New York, NY 10016-5991

Phone: (203) 702-7660
Fax: (203) 775-5177

www.aiche.org

Additional copies of this publication are available from:

Curran Associates, Inc.
57 Morehouse Lane
Red Hook, NY 12571 USA
Phone: 845-758-0400
Fax: 845-758-2634
Email: curran@proceedings.com
Web: www.proceedings.com

TABLE OF CONTENTS

EFFECTIVE APPLICATION OF PROJECT RISK MANAGEMENT

Stephen L. Cabano

Pathfinder, LLC
Cherry Hill, NJ

Biography

Steve Cabano is President of Pathfinder, LLC, a project management consulting firm specializing in the planning, development and execution of domestic and international capital plant construction projects. With more than 25 years of project engineering and project management experience in the industrial and commercial facilities industry, he has provided cost, scheduling, procurement, and similar project related services for owner and government clients in the Petroleum, Petrochemical, Chemical, Environmental, Pharmaceutical, Industrial and Commercial industries in the United States and throughout the world.

Introduction

In today's complex capital project environment, a project manager has a myriad of risks to consider and address. The traditional risks include: managing to the project cost and schedule commitments, addressing regulatory and environmental concerns, availability of internal resources vs. contracting out for services, among others. In today's increasingly active capital project execution environment, a whole new set of potential risks have arisen. Several of these include availability of competent engineering and construction contractors, experience of engineering and construction supervisors, material and equipment availability, the multi cultural nature of today workforce, etc.

Basis of Observations

This paper will address capital project risk management issues from an owner's perspective concerning projects in the process industry. For the purposes of this paper, we define the process industry as the oil, gas, chemical, petrochemical and related industry sectors. We will also discuss the application of risk management tools and techniques across an industry accepted, 5 Phase "Best Practice" Capital Project Delivery Process as outlined below in Figure 1:

EFFECTIVE APPLICATION OF PROJECT RISK MANAGEMENT

Figure 1
The Project Management Process

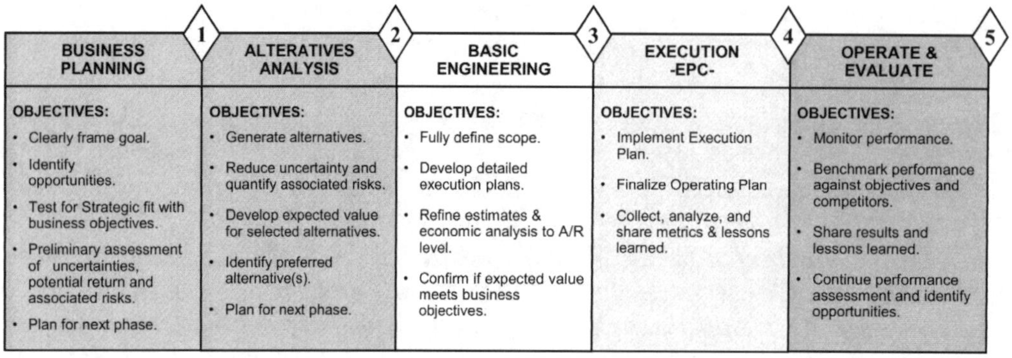

The Risk Management Process

The Project Management Institute (PMI) identifies Risk Management as one of the basic Project Management Bodies of Knowledge. (See Figure 2)

Figure 2

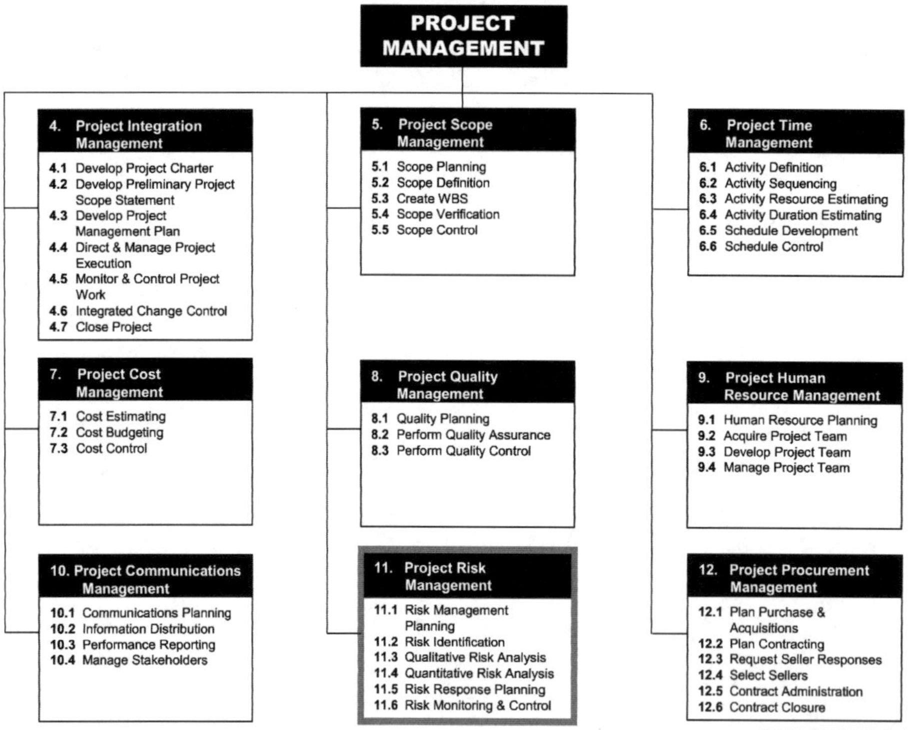

2008 PMBOK Guide

EFFECTIVE APPLICATION OF PROJECT RISK MANAGEMENT

PMI defines Risk Management in the following 6 categories:

- Risk Management Planning
- Risk Identification
- Qualitative Risk Analysis
- Quantitative Risk Analysis
- Risk Response Planning
- Risk Monitoring and Control

Let's take each one of these categories and identify the issues related to each in today's project execution environment.

Risk Management Planning

For a Risk Management approach to be effective there needs to be a company-wide established process which is applied, utilized and adhered to by all project representatives, as well as a project specific risk management plan integrated into the project's execution plan.

Company-wide processes should outline the corporate tools, resources, and deliverables by phase. A series of support "think lists"/risk considerations should be established for each project phase to assist the project team in identifying project risks. An example of this is outlined below in Figure 3:

Figure 3

Risk Management Think List by Project Phase

OPPORTUNITY (1)	ANALYSIS (2)	DEFINITION (3)	EXECUTION (4)	OPERATE (5)
• What are the business risks? • What market influences will impact this opportunity? • How will these risks impact the company?	• What are the opportunities and threats for each alternative evaluated? • How do these risks impact cost, schedule, operations, reliability, etc.? • What is the benefit of implementing a new technology? What are the risks?	• What are the project risks associated with the selected alternative? • What are the business risks associated with the selected opportunity? • How will the identified risks impact the overall project? • Are there mitigation options that should be incorporated into the base scope/execution plan?	• What new risks have been uncovered during detailed engineering, procurement, construction and Start-up? • How will these new risks impact the commitment made to management? • Are these late risks manageable and will they impact operations?	• What will impact the effective handover to operations? • What risks can be mitigated through application of effective Lessons Learned? • Are there risks that will affect operability/ maintainability and therefore operating costs and/or reliability/ utilization factors?

EFFECTIVE APPLICATION OF PROJECT RISK MANAGEMENT

The Project Execution Plan (PEP) should also have a specific Risk Management sub-section dedicated to each of the defined sections of the PEP. This sub-section should define the risks associated with that component of the PEP and describe how the team will handle the defined risks. The sections of a "Best Practices" PEP Include:

Section 1 - PROJECT CHARTER

- Strategic and Business Objectives
- Project Objectives
- Scope of Work
- Support of Local Communities/ Workforce/ Enterprises

Section 2 - PROJECT PLANNING BASIS

- **Risk Management Plan**
- Organization Plan
- Master Plan/ Schedule
- Contracting Plan
- Funding and Cost Estimate Plan

Section 3 - PLAN FOR IMPLEMENTATION

- Design Management and Value Improving Practices(VIP) Plan
- Constructability Plan
- Procurement and Materials Management Plan
- Construction Management and Labor Plan
- Mechanical Completion/Commissioning Through Start Up Plan
- Regulatory and Environmental Management Plan
- Project Controls Plan
- Quality Management Plan
- Safety Management Plan
- Informational Management and Communication Plan
- Document Management Plan
- Operations Management Plan
- Plan to Capture Lessons Learned
- Special Factors Plan (Modularization, Project Suspension/ Restart Plan etc.)

EFFECTIVE APPLICATION OF PROJECT RISK MANAGEMENT

In a Best Practices PEP, the risk sub-section includes a write-up on the Risk Management Plan, the identified risk, what impact the risk will have on the project (+/-), how the team intends to address the risk, who is responsible for implementing the mitigation approach, when the mitigation is intended to be completed and what benefits (quantifiable/tangible of cost/schedule improvements) should be expected as a result of implementing the recommended mitigation. This is often called the risk register and becomes part of the PEP, but is also updated. Risks are mitigated as other risks become apparent.

Risk Identification

As an initial step in the Risk process, the team needs to "know what it does not know". It needs to brainstorm project issues that can both negatively and positively impact the project and then determine how probable the risk occurrence will be, as well as the impact to the project if the risk occurs. This comes from experience. Having "been there, done that". It also requires research of Lessons Learned from previous projects (inside and outside of the company) which will, at the very least, stimulate thought provoking issues that the team can evaluate for possible applications to their project(s).

So, based on the above discussion of limited experienced resources across the industry, how do we know what we do not know? How do organizations today share the limited project experience across projects to gain the knowledge and input for all projects rather than targeting just a few projects?

A traditional step includes implementation of a Risk Identification workshop in coordination with the Project Execution Planning workshop. As the team defines the execution approach that best suits the project and business objectives, various risks would be unfurled and documented as part of the PEP. The risks would then be assigned to responsible parties. These parties, together with the project team, would define mitigation strategies for each of the identified risks. This approach works well when the project team is made up of highly experienced resources representing their respective disciplines. Today's teams do not necessarily have this cross section of experience and must rely on other risk identification techniques.

Today, we rely heavily on checklists of questions and Lessons Learned related to various aspects of the project scope and execution strategy. These checklists only go so far – as the individual asking the questions may not see the inherent risk in how these questions are answered.

There is one technique that has proven effective but takes a mature organization to assure appropriate implementation. What some organizations have realized is that, with the limited pool of experienced project management resources available, an effective approach to gaining the best use of their experience is not necessarily to put them on a single project, but to have them act as an independent review resource that all projects must go through at key junctures in the project. The experienced individuals are asked to "Put their black hats on" and see what issues they see in the scope definition, execution plan, etc. of the project as it moves through the companies. These individuals need to have the intestinal fortitude to drill down into details at key junctures in the project to better prepare the project team for the future project phases. This function is often part of a corporate-wide Project Management Organization (PMO), which is a shared service across the organization.

As projects become more internationally integrated and technologically complex, some of the consistent risks identified in today's project environment include:

- Limited experienced planning, oversight and execution resources
- Competing business demands
- Significant competition for limited pool of qualified Engineering and Construction contractor resources
- Inadequate scope and/or site condition definition
- Escalation variability
- Supplier/Fabricator shop loading (over-commitment impacting delivery schedules)
- Labor Productivity – dwindling supply of qualified craft workers
- Extreme sensitivity to economic conditions/ capital investment reticence
- Discipline to follow established strategies, plans, processes and procedures (schedule pressures)
- Ill-defined contractual requirements (scope of work and services)
- Insufficient recognition of host country requirements/project environment (work rules, labor availability/qualifications, etc.)
- Logistical challenges (world-wide procurement, customs/duties, transportation, infrastructure at location, etc.)
- Learning curve costs (just-in-time recruiting, on-boarding, orientation)
- Staff turnover
- "First time" consortium/"partnership"/"alliance" arrangements – integration, coordination, systems and procedure compatibility
- Others...

These are but a few examples of the risks and challenges facing business and project ventures, particularly in mega-projects and/or international/multi-national situations/endeavors.

EFFECTIVE APPLICATION OF PROJECT RISK MANAGEMENT

Risks today that are typically overlooked are often related to the execution environment more so than with the technology and processes involved. As stated previously, lack of experienced project management resources is a huge risk issue today. Availability of competent international suppliers that can deliver the quality and timeliness required is another major concern, as is availability of contractor teams that have worked together in the past and are not thrown together solely for the project at hand. Multi-cultural execution teams also represent some unique challenges (in some cases).

Qualitative Risk Analysis

Identifying project risk is only part of the battle; it is time and cost prohibitive to mitigate all possible project risks. With this in mind, the team must prioritize project risks and select the most effective Risk Response for each risk based on business and project goals and objectives. The prioritization process must include a measurement of the impact the risk will have, if it occurs, as well as the probability that the risk will occur. A typical risk matrix is included below as Figure 4:

Figure 4

Risk Matrix

Scope of use: Utilized to assess, evaluate and categorize risks associated with capital project planning and execution, and Major Projects Contracts, Purchase Orders.

Project Specific

Consequence (Note: For purpose of selecting impact of event, utilize most severe outcome of Project Specific and Corporate consequences analyzed)				Likelihood — How often would it be expected to have similar causes or circumstances aligned for the event with the defined consequences to re-occur? Please make reference to site, company and industry historical data to help predict the re-occurrence frequency of such an event.				
Project Specific				Event will almost surely not happen	Event will quite unlikely happen	Event will quite likely happen	Event will very likely happen	Event will almost surely happen
Cost/Financial Impact	Schedule	Scope	Severity	**Very Low** — < 5%	**Low** — =>5% and < 20%	**Medium** =>20% & < 50%	**High** — =>50% & < 80%	**Very High** — =>80%
Cost Δ > $10MM or Cost Δ > 5%*Capex; which ever is smaller	> 6 months	Scope or Quality does not meet performance or business expectations	Very High	Moderate	Significant	Significant	Critical	Critical
1MM <Cost Δ ≤10MM; or 2%*Capex< Cost Δ ≤ 5%*Capex; which ever is smaller	> 3 months but ≤ 6 months	Scope or Quality unacceptable to Project Manager	High	Negligible	Moderate	Significant	Significant	Critical
100K <Cost Δ ≤1MM; or 1%*Capex< Cost Δ ≤ 2%*Capex; which ever is smaller	> 1 month but ≤3 months	Significant impact on scope or quality of major project deliverable and/or component	Medium	Negligible	Moderate	Moderate	Significant	Significant
10K <Cost Δ ≤100K; or 0.5%*Capex< Cost Δ ≤ 1%*Capex; which ever is smaller	> 1 Week but ≤ 1 month	Impact on scope or quality of minor project deliverable and/or component	Low	Negligible	Negligible	Moderate	Moderate	Significant
0 <Cost Δ ≤10K; or 0%*Capex< Cost Δ ≤ 0.5%*Capex; which ever is smaller	Negligible	No noticeable scope or quality impact	Very Low	Negligible	Negligible	Negligible	Negligible	Moderate

Consequences

EFFECTIVE APPLICATION OF PROJECT RISK MANAGEMENT

This too is a function where the above defined shared PMO risk function can be of benefit. The experienced project representative can bring some realism to the quantitative assessment of the identified risks. Inexperience can lead to downplaying what might seem to be minor issues, which if thoroughly analyzed, could have major impact to project success potential.

Quantitative Risk Analysis

The team must also measure the impact a risk might have on the project cost and schedule. The main issue here is that one risk may impact a number of project activities. As an example, site logistics or poor transportation to and from the site which might include limited road accessibility, multiple bridges and overpasses, poor road conditions, etc., will not only impact delivery of equipment and materials but also may cause construction productivity issues (people cannot get to the site in a timely manner), emergency response timing, local pubic relations issues, etc. These impacts all need to be taken into consideration when quantitatively measuring the impact of an identified risk.

There are a number of excellent risk analysis tools on the market that, when used properly, can assist the project team in measuring the impact risks have on the project cost and schedule. We will not go into their functionality in this paper, but many use a Monte Carlo simulation approach to develop risk profiles for both cost and schedule. This type of analysis assists the team in selecting which Risk Response effort provides the most project/business benefit for the investment. The project team needs to be well-versed in utilizing these tools, as they can lead to a misunderstanding of the impacts of risk if applied incorrectly. Owner management also needs to define what level of cost and schedule probability the project must "shoot for", as this will affect how much time and cost contingency are applied to the project.

If a selected risk mitigation strategy is applied, the cost, time and resources required to implement this strategy must also be considered, quantified and included in the base cost estimate and schedule. This is often overlooked and leads to false risk analysis readings.

Risk Response Planning

Probably the most important aspect of Risk Management is what the team will do to address the risk. Identification and measurement are critical but the real test is in how the team addresses the risks. There are a number of risk response options which are outlined in Figure 5 below:

Figure 5

Responding to Risk

RESPONSE TYPE	EXAMPLES
Avoidance	Choosing only high probability options, eliminating high risk approaches on parallel programs, screening subcontractors.
Prevention	Focusing on preventive measures, extensive potential problem analyses, extensive bench testing, mockup, and simulation work
Reduction of Impact /Mitigation	Focus on contingency planning, large contingency funding, frequent progress reviews
Compromising	Tending toward low risk options, choosing low return options, pursuing redundant paths
Transfer	Through subcontracting, warranties, performance bonds, insurance
Assumption	Any decision to proceed implies assumption of risk

Often risk registers are developed but not implemented; risk responses are identified, but too late. Many times the symptom is addressed but not the root cause (example: poor productivity in the field - the owner authorizes more resources but the true causes were engineering quality and/or late material deliveries).

In the case of engineering quality, the risk may be avoided by the selection of high caliber technical personnel within contractual terms.

Risk Monitoring and Control

Once the Risk Response approach has been implemented, management and project team representatives need to be assured that the selected response is doing what it intended to do. Project Control plays a key role here. The team needs to be assured that the money, resources and time allocated to a risk mitigation option are successfully achieving the benefits expected. If not, the team needs to reevaluate the option and develop an alternate approach.

One of the main issues we have today is the weakness in project controls across the industry. Of all the 'brain drain' suffered across the project disciplines, the degradation of the project control capability has been the greatest. This is a risk in itself. The team needs to be assured that it has a strong in-house, project control capability to assure the team can measure the effectiveness of the mitigations options.

EFFECTIVE APPLICATION OF PROJECT RISK MANAGEMENT

Why Is Risk Management so Important Today?

Today's projects are more complex, involve more planning and execution entities, are executed in a more challenging environment and involve multi-national resources. These issues, combined with a less experienced workforce are a recipe for disaster. Identification, measurement and response planning can aid in eliminating and/or minimizing the negative impact that risks may have on project success.

Who is responsible?

Risks are shared by the owner, contractors, suppliers, and other project stakeholders. Ultimately, the owner bears the burden of accountability for all project risks. At the end of the day, it's the owner's money. Product delivery is an owner issue (especially if they have made sales agreements and commitments to their customers). Quality becomes an owner issue – as the owner has to operate the facility for the next 20 – 30 years.

In Conclusion

A Project Manager in today's world cannot effectively manage a project without also managing the project risks. How aggressively one tackles the risk is dependent on the business/project drivers for the project, scope definition, execution team experience, etc., and is all part of the project execution strategy defined in the planning process. Risk management is one of the most critical skill sets a Project Manager needs to successfully manage projects in today's environment. Mastering this knowledge area goes a long way toward assuring achievement of project goals, objectives and success factor requirements.

(34b) The Legal System As a Driver for Innovation In Energy Technology

Richard D. Jordan, IP, Dickinson Wright PLLC, Washington, DC and Mary Ternes, McAfee& Taft, Oklahoma City, OK

This presentation will provide an overview of how the legal system acts as a primary driver for innovation in energy related technology, considerations for funding from public and private sources, and considerations for determining how to best protect intellectual property developed in the innovation effort. The presentation will be made by invited policy and legal experts. These experts will discuss policy and procedural considerations in which governmental agencies, such as the Environmental Protection Agency, "intervene" for the good of society to require industry to provide "cleaner" technologies. These legal drivers in turn cause the industry to invest a great amount of funds to innovate in order to meet the new legal requirements. Panel experts will discuss: 1) a few of the policy issues that are considered by governmental institutions in implementing the new laws and regulations, 2) government grants or funds that are made available to stimulate innovation in such a capital intensive market, and 3) considerations on how to protect valuable intellectual property that will be created in getting the new innovations to market.

See more of this Session: Law, Risk, and Ethics of Technology and Project Management
See more of this Group/Topical: Management Division

(34d) Patenting Basics

Vivek Koppikar, Patent Attorney, Dale City, VA

This partially invited session will provide a fundamental review of the patent procurement process and patent law in general, and is designed for anyone who has any type of interest in the patenting process and/or deals with patents in their work in industry and/or academia. Little or no background knowledge in patents is needed by those intending to attend this session. The session will begin with the very basics of the patenting process as well as providing them with basic terminology. Next, the session will provide a basic synopsis of the patent procurement process with the US Patent and Trademark Office. The session will then focus on portions of a patent most relevant in patent disputes and patent litigation. Case studies involving various patent scenarios will be presented. The session will wrap up with a question and answer session where attendants can ask basic questions dealing with the patent procurement process and patent law in general. This session will provide a foundation for two following patent sessions focusing first on drivers for innovation in energy technology and, next, on litigation for engineers. Abstracts are solicited for case studies from lawyers and engineers with various patent experience

See more of this Session: Law, Risk, and Ethics of Technology and Project Management
See more of this Group/Topical: Management Division

Ethical Project Management

Dr. Eldon R. Larsen
Marshall University

ABSTRACT

Project managers are often faced with situations that present ethical dilemmas. These situations can be anything from very simple, like something in a conversation, to something very complex, like how to accept or not accept a gift. This paper presents some principles and suggestions regarding ethics in project management. It also highlights the main concepts in the Project Management Institute's ® Code of Ethics and Professional Conduct.

INTRODUCTION

Many people have written about ethics. The objective of this paper is not to re-hash past studies. Rather, I attempt to present significant concepts and principles through both quotations and discussion that will help the reader gain a vision of what it means to be ethical in one's behavior. This applies very much to project managers, as they are frequently making decisions that have ethical questions surrounding them.

CARDINAL RULES OF ETHICAL BEHAVIOR

Following are a few quotations in which are embedded basic principles and rules of ethical behavior. For several of the quotations are given "sub-quotations" that are closely related, but are expressed in entirely different ways.

- "Therefore all things whatsoever ye would that men should do to you, do ye even so to them" (Matthew 7:12)
    - "You can easily judge the character of a man by how he treats those who can do nothing for him." (James D. Miles)
- "And above all things have fervent charity among yourselves: for charity shall cover the multitude of sins." (I Peter 4:8)
    - "When I do good, I feel good. When I do bad, I feel bad. That's my religion." (Abraham Lincoln)
- "Thou shalt love thy neighbor as thyself." (Leviticus 19:18)
    - "Laws control the lesser man. Right controls the greater one." (Chinese proverb)
- "To thine own self be true, and it must follow, as the night the day, thou canst not then be false to any man." (William Shakespeare, *Hamlet)*

- "To know what is right and not do it is the worst cowardice." (Confucius)

FOOTNOTES TO THE CARDINAL RULES

- Though you legally can do something to someone else, it might not be alright to do it to him.
- Ethical people normally follow the laws, but are not constrained from going beyond the bare minimum of the law to do what is right
- Good ethical behavior is at the heart of excellent leadership—you do what is right because it is right, not because it benefits you monetarily or otherwise
    - "Keep true, never be ashamed of doing right; decide on what you think is right and stick to it." (*George Eliot)*
- "If moral behavior were simply following rules, we could program a computer to be moral." (Samuel P. Ginder)

ETHICAL PROJECT MANAGEMENT AND PRINCIPLES

Ethical project management involves right behavior personally, legally, socially, and environmentally. Inconsistencies in one area can affect behavior in another area. Consistency of ethical behavior in all areas strengthens a person's integrity.

We normally consider the goal of project management to be to achieve project objectives. Some questions we should ask ourselves, however, are the following:

- How well do we meet the objectives, and within what constraints?
- Does it matter how we get there?
- Are some objectives more important than others?
- How do we treat team members? Stakeholders?

How might ethical considerations affect the answers to the above questions?

Many organizations have a code of ethics by which they expect their participants to abide. However, it is important to consider how the code was developed. I suggest that you should always questions a code of ethics which is based solely upon the premise that right or wrong is determined either by opinion or by statute. Truth is true whether people agree with it or not. And even making a statute does not change that truth is still true, and false is false. True principles will always be true, no matter how hard we might try to ignore them, change them, or go against them.

The fact that true principles will always be true can present us with a *principles dilemma*. One of the challenges we face is when principles seem to conflict with each other. For example, is it right to tell a lie in order to save an innocent person's life? Many people would probably say "Yes." This leads us to realize that one principle might be more important than another, even though they are both true (Don't kill. Don't tell a lie.). It is often our conscience and our experience that tell us what to do in these circumstances. In fact, good ethical behavior is closely tied to a sensitive conscience.

PRINCIPLES AND VALUES

Principles and values are not exactly the same thing, although they can be very closely related. Most values are built upon some principles. Values are like roadmaps that guide us in our decision-making processes. Therefore, it is important for us to identify and understand our values—companies and organizations need to understand their own core values. Once we better understand our values, we can more easily make an ethical decision through some kind of structured analysis.

Cultural norms can affect what is considered acceptable or not. While principles (truth) will always be true, the application of the principles might change significantly, depending upon the situation. Values (how we perceive life) sometimes vary with culture, religion, upbringing, etc., and therefore change the situation. They relate to perspective. So, it is important to understand cultures when considering what decision we face is ethical. We don't have to break principles when we adjust for different values—they are not the same thing.

Project management ethical considerations often involve the following questions.

- Are you obeying the law? Are you following guidelines provided by your organization and the customer's organization? Which one should take priority?
- Are you treating other people the way you would want to be treated, and would you do it either in private or in public and still have personal peace?
- What is driving your actions? What if everyone did what you are considering doing? Would that be good?
- Is your choice the right thing to do, according to the principles you know to be true?

There are also some legal considerations when dealing with ethical questions. For example:

- "The Foreign Corrupt Practices Act – FCPA, 1977, states with respect to bribery that it is a crime for a bribe to be made to a foreign official or political party for the purpose of obtaining or retaining business or for directing business to another person even if this flows through an intermediary or consultant.
- "However, not all payments are defined as "bribes." Payments for "routine" government action are allowed under the FCPA including the dispensing of permits, licenses and police protection as noted in the example."[1]

Following is an ethical, legal situation to illustrate some of the challenges in making ethical decisions:

- Country A allows companies to perform drug testing on employees, but allow employees to refuse to be tested
- In Country B, companies can mandate that employees and contractors submit to drug testing, and refusal can be grounds for dismissal
- A Country A contractor is asked to perform work for a company in Country B, that asks the contractor to submit to a drug test. He refuses. Right choice or not?

Even though the contractor has the right to refuse drug testing according to the policies of his own home country A, because he is doing work for company in country B he should follow the more stringent policies of country B.

The Project Management Institute's Code of Ethics and Professional Conduct

The Project Management Institute's Code of Ethics and Professional Conduct[2] includes many fine points of ethical behavior for project managers. For this reason, it is useful to summarize them here.

1. Take responsibility for the decisions we make or don't make, and for the resulting consequences
2. When making decisions, take into account the best interests of society, public safety, and the environment
3. Keep commitments
4. If we make a mistake, take ownership and responsibility, and correct it. If others make mistakes, communicate appropriately as quickly as possible
5. Protect proprietary or confidential information
6. Be informed regarding policies, laws, regulations related to our work, profession, and volunteer activities
7. When substantiated by facts, report unethical or illegal conduct to appropriate management and, if necessary, to those affected by the conduct
8. Be informed about the customs of others and avoid being disrespectful
9. Listen to other people's opinions and views and try to understand them
10. When a conflict occurs, deal directly with the people involved
11. Be professional, even if this conduct is not reciprocated
12. Negotiate in good faith
13. Do not use power to benefit personally at the expense of others
14. Do not act abusively
15. Respect the rights of others

16. Demonstrate transparency in decision making
17. Be impartial as much as possible
18. Allow equal access to all who are authorized to have information
19. Make opportunities equally available to qualified candidates
20. Actively and fully disclose potential conflicts of interest to appropriate stakeholders
21. Refrain from engaging in decisions or influencing others when we have a conflict of interest, unless or until: we have made full disclosure to affected parties; we have an approved mitigation plan; and we have obtained consent from affected stakeholders to proceed
22. Do not hire/fire, reward/punish, or award/deny contracts based on personal considerations like favoritism, nepotism, bribery, etc.
23. Do not discriminate against other people based on things they cannot control or based on their religion
24. Apply your organization's rules without favoritism or prejudice
25. Seek the truth
26. Communicate truthfully
27. Provide timely, accurate information
28. Make implied or explicit commitments in good faith
29. Strive to create a non-threatening environment for people to tell the truth

SUMMARY

Ethical project management involves many facets of application, all of which should be based upon foundational principles of good management and human relations. The basic rules or principles of ethical behavior have existed for thousands of years and will continue to exist for thousands more. Truly successful project management applies these ethical principles fairly according to the situation and the needs of the people involved.

REFERENCES

1. http://www.bia.ca/articles/pj-project-managers-ethical-dilemma.htm
2. http://www.pmi.org

(61a) The Future for 3M in Nanomaterials Science and Engineering

Larry A. Wendling, Corporate Research Laboratory, 3M, St. Paul, MN

3M's fundamental business model can be described as technology-driven innovation. Having 45 technology platforms, 3M ranks among industry's R&D leaders in terms of its breadth of technical capabilities. The ability to access novel size dependent properties via nanoscale materials science and process engineering potentially impacts the majority of all our technology platforms and provides significant opportunities for differentiated product performance across many of 3M's businesses. Recent advances in nanomaterials science and engineering at 3M providing novel optical, mechanical, electrical, and catalytic properties will be described.

See more of this Session: 2011 Annual Meeting Plenary - Chemical Science Innovation: Future of the US Chemical Enterprise I (co-sponsored by the Council for Chemical Research)
See more of this Group/Topical: Liaison Functions

(61b) Educating Global Chemical Engineers in the 21st Century

Robert A. Brown, Boston University, Boston, MA

Chemical engineering thrived in the United States during the 20th century because of the confluence of discoveries in chemistry, innovations in process technology, the abundance of raw materials, and markets fueled by economic growth. What will chemical engineering in the United States look like in 2050? Will we experience the same level of innovation and prosperity in an economy that is driven by substantial growth off our shores? This presentation will discuss the impact on the chemical engineering discipline of the emergence of China, India and other countries as economic equals. Some potential strategies will be offered for professional and institutional leadership in the globalized economy that lies ahead.

See more of this Session: 2011 Annual Meeting Plenary - Chemical Science Innovation: Future of the US Chemical Enterprise I (co-sponsored by the Council for Chemical Research)
See more of this Group/Topical: Liaison Functions

(61c) Novel Structures for Public-Private Partnerships: Case Studies with Medical Devices and Biopharmaceuticals

Sangtae Kim, Morgridge Institute for Research, Madison, WI

At a time when the growth rate in research opportunities towers over the funding capacity of traditional sources of sponsorship, universities are turning to translational research and the associated licensing revenue streams as the untapped resource to close the gap. This hope is challenged by the historical track record which shows that most universities do not have a successful track record in technology transfer. In this context, this presentation examines the current landscape for technology transfer via spin-off companies generated from university research. The classical view of a "basic research phase" and a "VC-supported development & commercialization phase" separated by the so called "valley of death" albeit simplistic, offers some insights into the high failure rates. The presentation proposes a new model for a public-private partnership with the goal of higher success rates for translational research and development. The model will be illustrated with recent case studies involving medical devices and biopharmaceuticals.

See more of this Session: 2011 Annual Meeting Plenary - Chemical Science Innovation: Future of the US Chemical Enterprise I (co-sponsored by the Council for Chemical Research)
See more of this Group/Topical: Liaison Functions

Corning's R&D / Innovation Approach

October 17, 2011
American Institute of Chemical Engineers

Joe Miller, Ph.D.
Executive Vice President
Chief Technology Officer

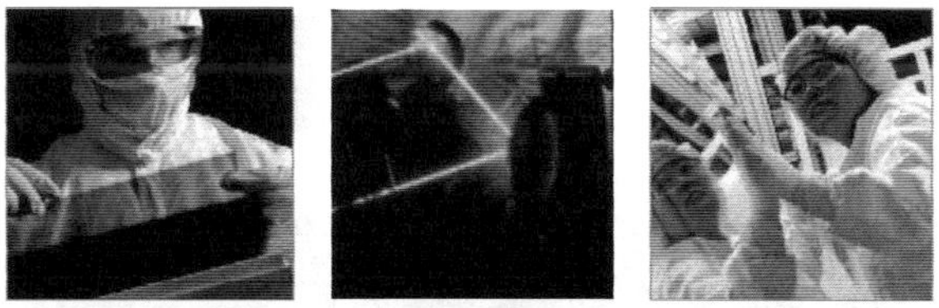

Corporate Snapshot

CORNING

2

Corning Incorporated
Who We Are

- Founded: 1851
- R&D Lab Established: 1908
- Headquarters: Corning, NY
- Employees: ~26,000 globally
- Sales: ~$6.6 billion
- Sales (incl. pro rata JV sales): ~$11 billion
- RD&E @ 9% of sales $670 million

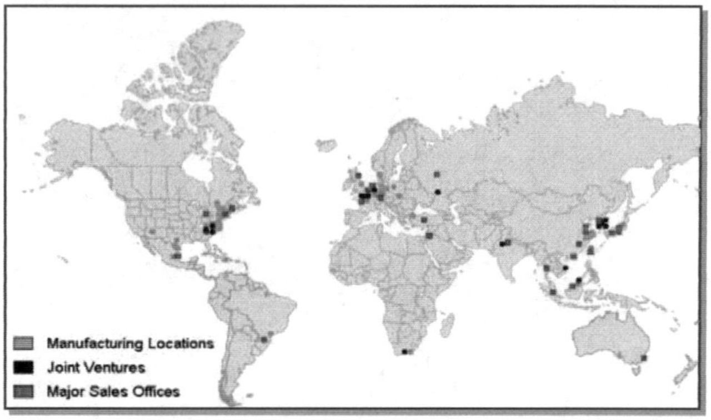

- Corning's global operations:

  ~ 65% of revenues from outside North America

  More than 50% of our employees are outside the United States

Corning Incorporated
Who We Are

Market Segments

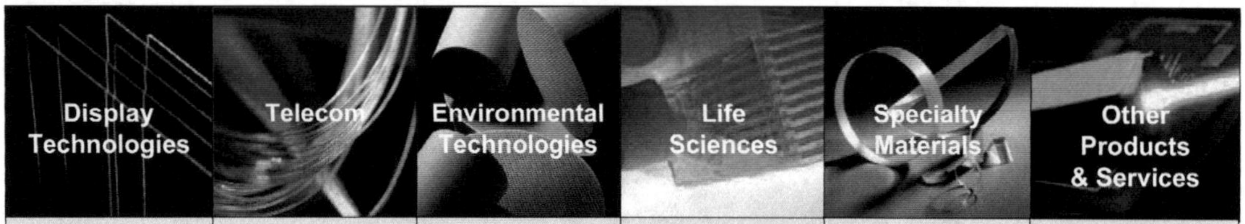

Display Technologies	Telecom	Environmental Technologies	Life Sciences	Specialty Materials	Other Products & Services
• LCD Glass Substrates • LTPS-LCD Glass Substrates	• Optical Fiber & Cable • Hardware & Equipment	• Emissions Control Products – Automotive – Diesel – Stationary	• Cell Culture & Bioprocess • General Laboratory Products	• Advanced Optics & Materials • Display Optics & Components • Semiconductor Optics & Components • Aerospace • Astronomy • Optical Metrology • Ophthalmic • Telecom Components • Specialty Glass	• Display Futures • New Business Development • Drug Discovery Technology • Equity Companies – Dow Corning Corp. – Samsung Corning Precision Glass Company, Ltd – Eurokera, S.N.C

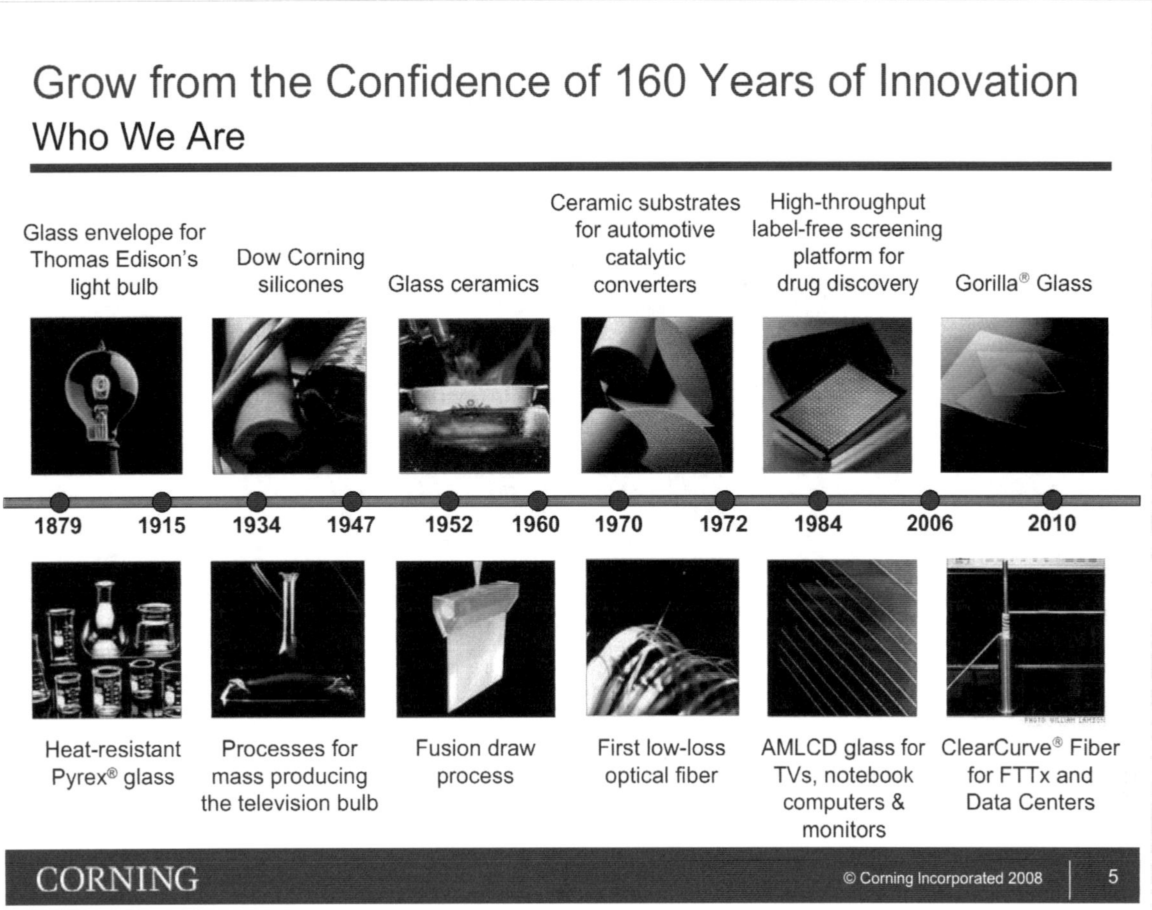
Grow from the Confidence of 160 Years of Innovation
Who We Are
Glass envelope for Thomas Edison's light bulb
Dow Corning silicones
Glass ceramics
Ceramic substrates for automotive catalytic converters
High-throughput label-free screening platform for drug discovery
Gorilla® Glass
1879
1915
1934
1947
1952
1960
1970
1972
1984
2006
2010
Heat-resistant Pyrex® glass
Processes for mass producing the television bulb
Fusion draw process
First low-loss optical fiber
AMLCD glass for TVs, notebook computers & monitors
ClearCurve® Fiber for FTTx and Data Centers
CORNING
© Corning Incorporated 2008
5

Economic Disruptions
Evolution of Strategy
CORNING
© Corning Incorporated 2008
6

Look where we started 8 years ago
Where have we come from?

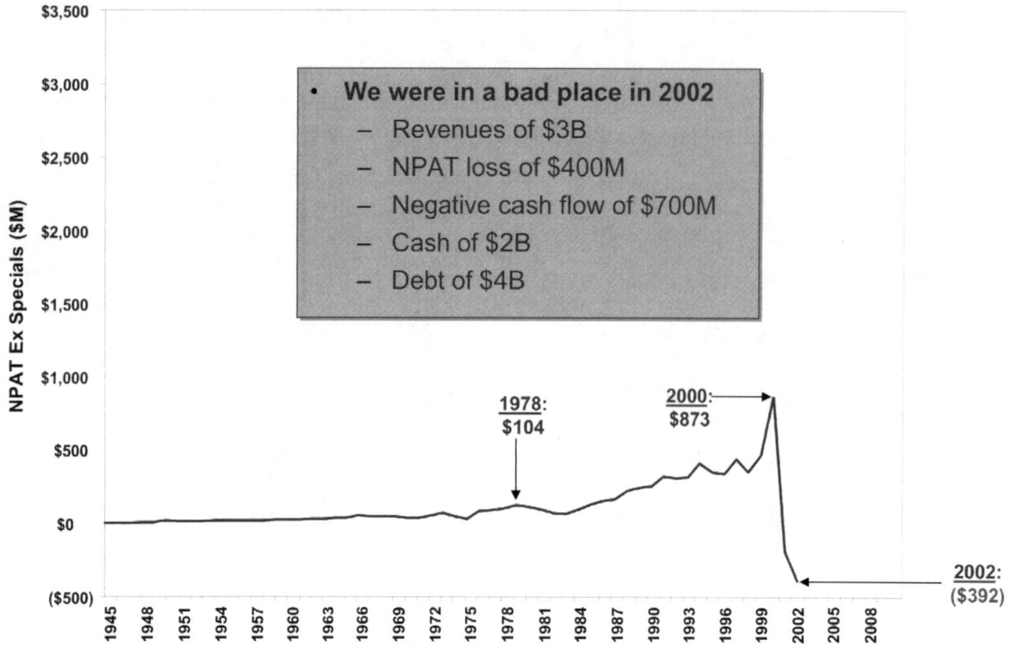

Corporate Strategy Framework
Another 150 Years of Innovation & Independence

We Need To Grow	***We Also Need Stability***
Grow Through Global Innovation	**Provide Stability and Balance**
Synergy	Tension

We Must Preserve Trust

Live the Values

Restructured and Aligned RD&E with Strategy Framework

- Eliminated worldwide network of satellite labs
- Concentrated R&D in Corning, NY, and Fontainebleau, France
- Reduced total RD&E spending by 50%
- Focused on strengthening core capabilities
  - Glass
  - Ceramics
  - Optical Physics
  - Process Engineering
- Added new competencies carefully – ones that added uniqueness when combined with our materials and process core
- Concentrated on supporting the growth of ongoing businesses
- Created an early stage business group to prime the innovation pipeline
- Added new innovation governance system

... and then began to reinvest in growth through innovation

Where have we come from?

We've delivered outstanding financial performance

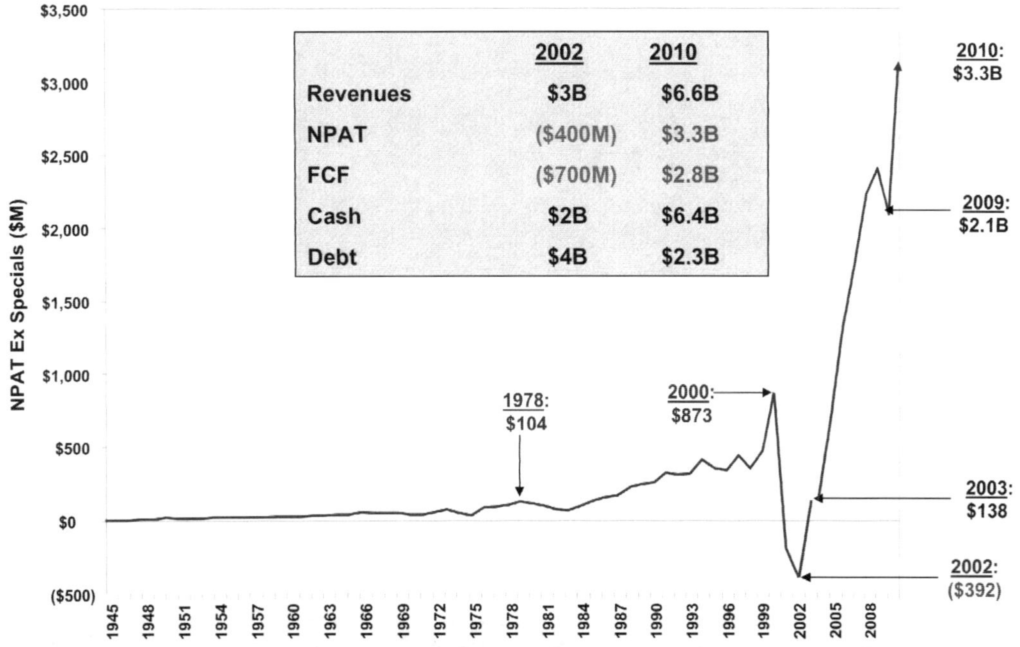

	2002	2010
Revenues	$3B	$6.6B
NPAT	($400M)	$3.3B
FCF	($700M)	$2.8B
Cash	$2B	$6.4B
Debt	$4B	$2.3B

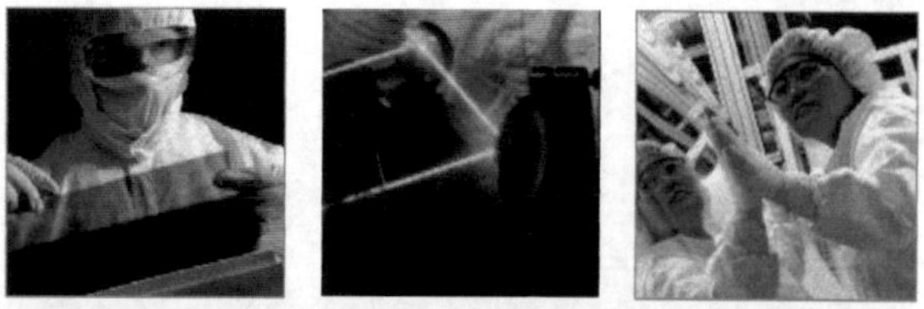

Strategy Implementation
The Innovation Recipe

Corning Has A Distinctive Innovation Recipe

- Take on tough problems
- Stick to our competencies
- Achieve dominant design
- Build strong IP barriers
- Capacitize quickly
- Sell to concentrated markets with high seller power
- Multiple JV's for access
- Significant RD&E and capital investments

Created one to two businesses/decade

Innovation Recipe Yields Keystone Components

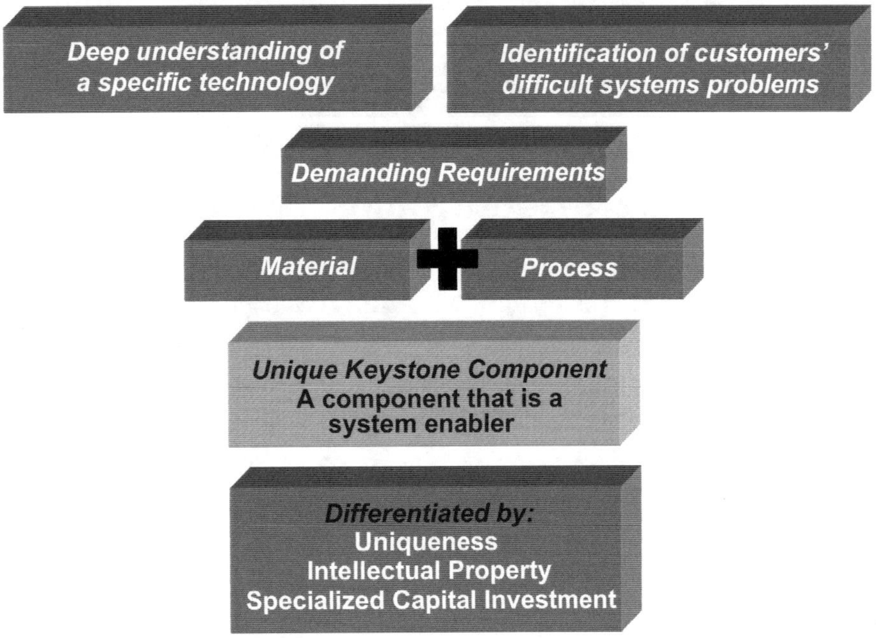

Keystone Components
Based on Integration of Multiple Competencies

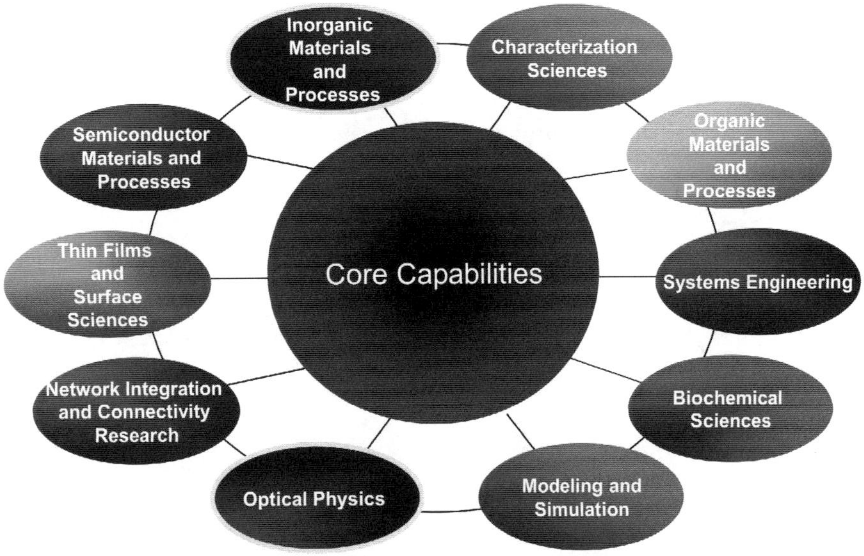

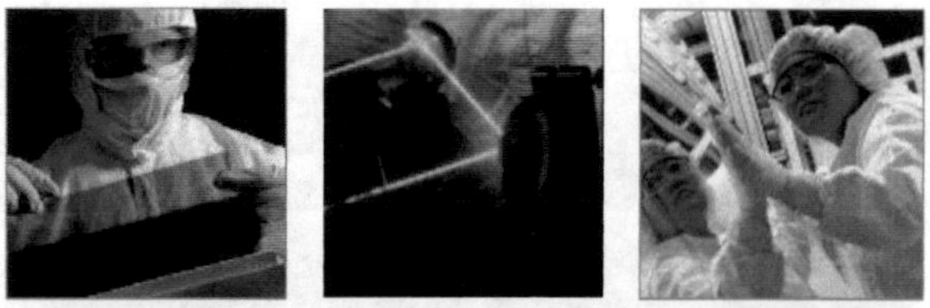

Innovation Strategy
Response to 2001

We consciously centralized our Research & Development activities and facilities

Central R&D Facility at Sullivan Park – Corning, NY

Sullivan Park Connected Globally in Europe, Asia and the U.S. West Coast

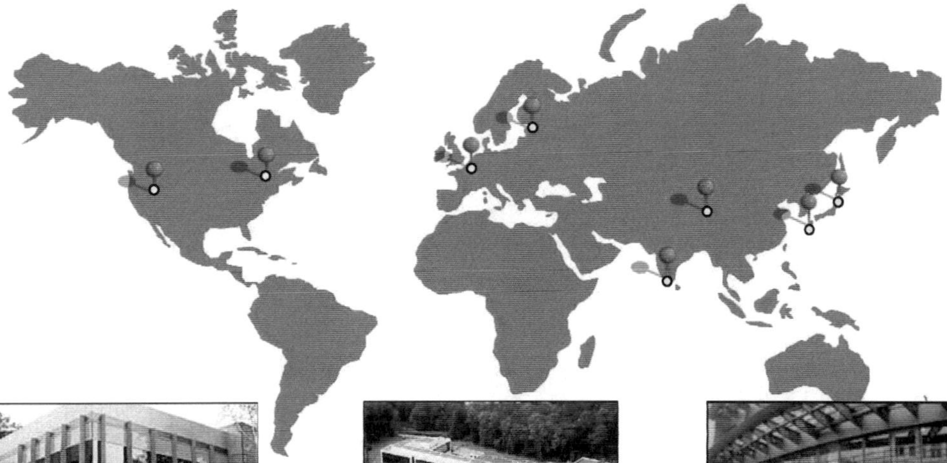

Corning West Technology Center
Silicon Valley, California

Corning European Technology Center
Fontainebleau, France

Corning Advanced Technology Center
Taipei, Taiwan

Hardened our Core

Strategic Approach to Competencies

Dig Deeper in Historic Competencies

- Precision melting, fusion, extrusion, and draw
- Glass, glass ceramics, and cellular ceramics
- Optical physics

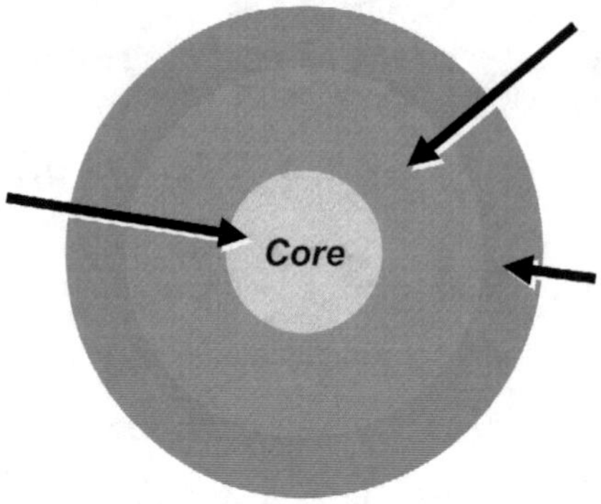

Add New Competencies

- Optical Semiconductors
- Biophotonics
- Carbon

Enabling Capabilities

- Process and materials modeling
- Organics
- Characterization
- Finishing
- Coating
- Systems engineering
- Assay development

Created New Corporate Innovation Governance

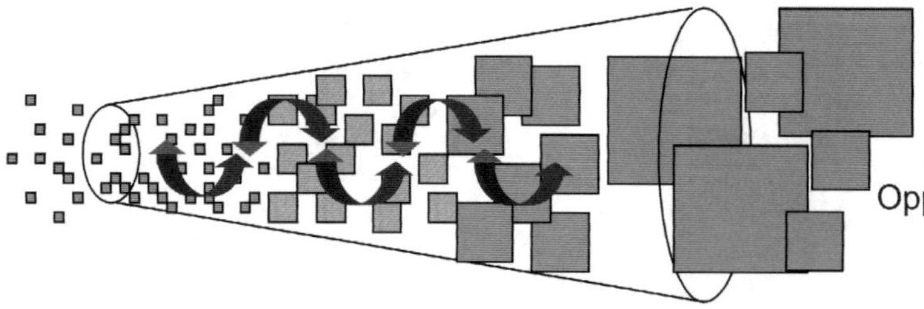

Corporate Technology Council

- **Create-the-Next Wave Portfolio**
- **Early stage opportunities**
- **CTO + Research & NBD Leaders**

Growth Execution Council

- **Invest-in-the-Future Portfolio**
- **Sort, pace and execute**
- **CEO, CTO + CTO Staff**

Technology & Strategy with the Board

Reinforced Five Stage Innovation Process

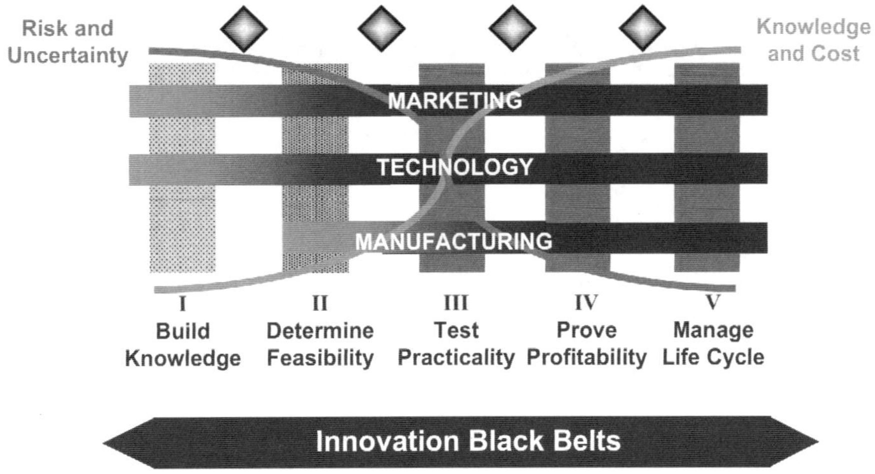

Research & Development Today

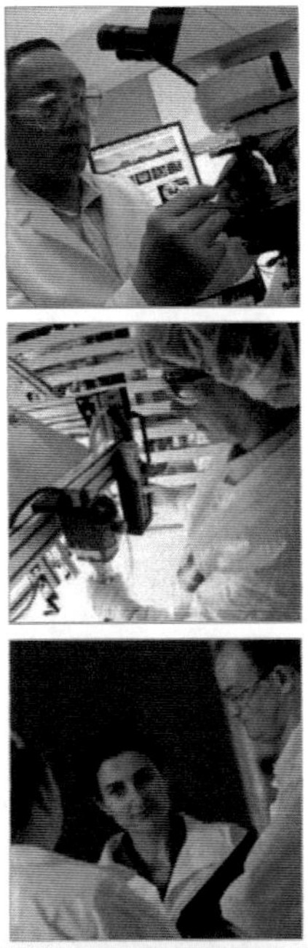

- RD&E investment intensity ~ 9% of sales
  - 5% Development & Engineering
  - 4% Research & New Business Development
- Core Competence R&D Model
- Centralized R&D Laboratory Model
  - Sullivan Park R&D Center, Corning, New York
  - 1900 employees (technical + support)
  - 40 different countries represented
- Global reach-out in Europe and Asia

Effective Technology Delivery to Current Businesses

Display	Diesel	Auto	Telecom	Specialty	Life Science
• LCD Substrates • Poly-Silicon Substrates • Thin Substrates	• Light Duty Diesel Filters • Heavy Duty Diesel • Stati…	• Next Gen Substrates 600/2 400/3 • Thin Wa…	• FTTx – Fiber, Cable, H&E • ClearC… • …ltra Low - loss Fiber • Pretium® EDGE - Data Center	• Gorilla… …for KrF • Sensor Products • Precision Optics	…stem Assays • Advanced Cell Growth Surfaces

~Two New Products Each Year for Each Business

CORNING 23

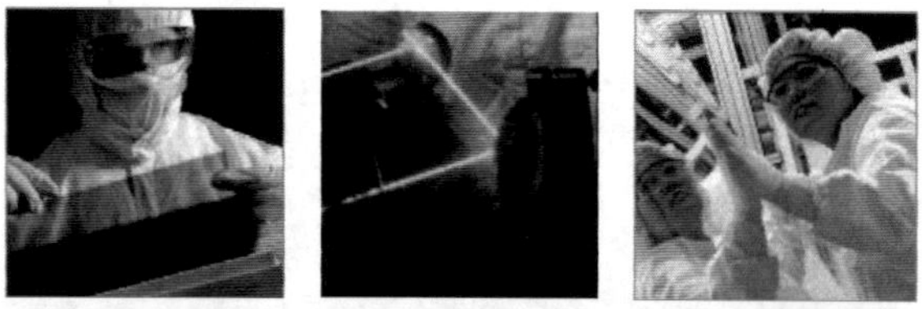

Innovation Strategy Response to 2009

CORNING 24

Adjacencies - The Third Strategic Leg

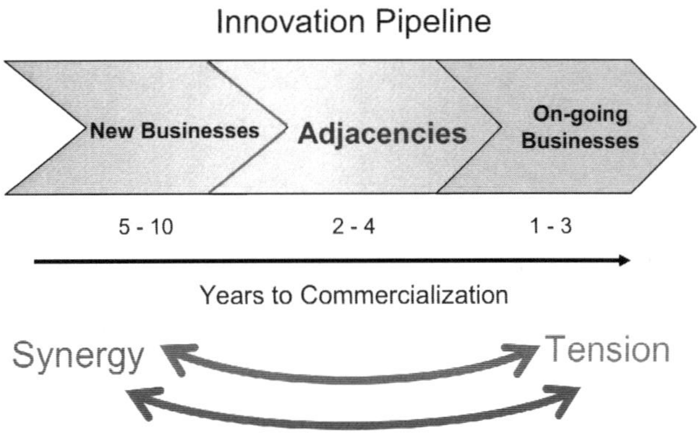

Forced the new leg into the middle

CORNING

25

With adjacencies, we tuned our strategy to create more balance in the portfolio using existing technology assets

- Specialty glass is our core materials and process capability
- We have substantial glass sheet forming process technology and manufacturing assets to support the Display Technologies business
- We redeployed some of those assets to address adjacent market segments in Consumer Electronics with thin, strong, damage-resistant glasses
- Here's an illustration of the adjacency concept

CORNING

26

Illustrative

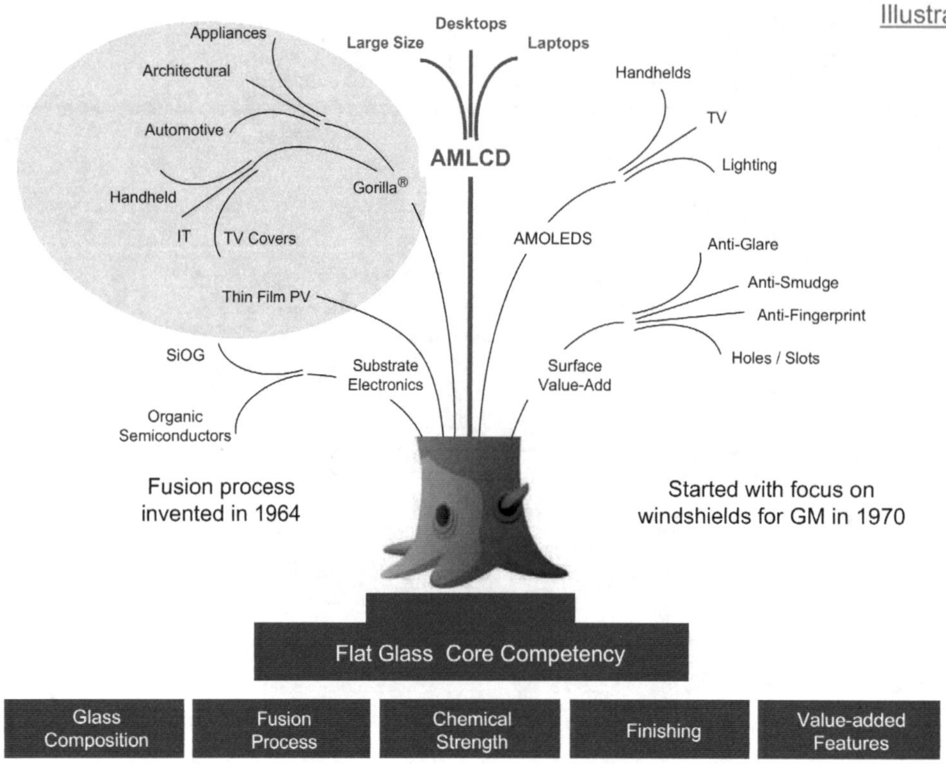

2011 Strategic Frame

- Corporate technology strategy has three legs
  - Support the ongoing businesses near term
  - Build an adjacency opportunities portfolio medium term
  - Maintain exploratory activities and new business development longer term
- Make room for significant shift to adjacencies
  - Resource the fastest growing segment – glass
  - Restructure New Business Development
    - Prune the New Business Development project portfolio
    - Direct some of those resources to adjacencies
  - Maintain exploratory research to create a pipeline of opportunities

2011 Innovation Portfolio Investment Allocation

Near Term	Medium Term	Long Term
60%	25%	15%
1-3 years	2-4 years	5-10 years

Near Term: Existing Businesses
- Display: Gen 10
- Diesel: AT/AC/CO
- Thin Film PV

Long Term: Reset in Stage I
- Mercury Abatement

New Opportunities
- Native Green Lasers
- Advanced Batteries
- CO_2 Sorption

2011 Innovation Portfolio

Near Term (60%)	Medium Term (25%)	Long Term (15%)
Major Businesses • Display • Diesel • FTTx & Next-Gen Data Center **Flat Glass Adjacencies** • Gorilla® Glass - Handhelds/IT Covers - Touch Screens - TV Covers • Thin-Film Photovoltaics	Advanced Life Sciences Advanced Consumer Electronics and Industrial Applications Advanced Wire/Wireless Optical Communications Energy & Environmental	• Organic Semiconductors • Thermoelectrics • Advanced Batteries • CO_2 Sorption • Ceramic Adjacencies • Optical Adjacencies
1-3 years	2-4 years	5-10 years

Gorilla® Glass Applications
Touch Devices

- "Projected capacitive" touch devices are changing the game
  - Multi-touch
  - More complex applications and gestures
  - Better optical performance

- Touch device growth accelerating
  - Handheld 20% CAGR
  - Notebook/netbook 60% CAGR

- Potential for sensor integration on cover glass

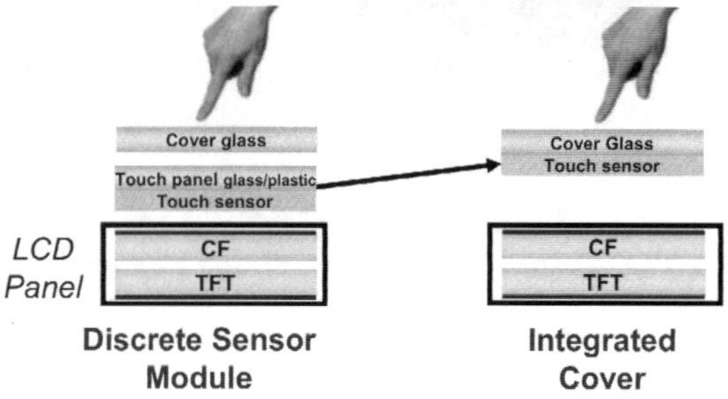

Gorilla® Glass - The Slate Effect

- Glass will play a significant role in this space
  - Thinner LCD glass
  - Thinner Touch sensor glass
  - Thinner, yet stronger cover glass will be required

- Weight and area increase 5 to 7x on slates vs. handheld devices

- System designers faced with significant challenges

- Gorilla a perfect fit for reliability

Glass Adjacency Innovation Results

- Gorilla® Glass is exceeding growth expectations
- Growing awareness glass as a new design element
- More than 30 major brands using Gorilla® Glass
- Next generation technology delivering significant improvements in both performance, capabilities and product possibilities
- With a potential to achieve close to $ 1B revenues in 2011

CORNING 33

Strategic Response Delivered Strong Technology Performance

- Technology leader in each of our market segments
- More than 50% of revenues over last three years from new products
- Corning leads the industrial materials segment in intellectual property performance
  - > 200 patents granted per year
  - 400 – 500 patent applications in each of the past two years

CORNING 34

Innovation Wisdom
What I've Learned

Dynamics of Innovation – Observations

- Innovation is more than a process
- Innovation is an interconnected system
  - Linked to strategy
  - *Supported* by process
  - Directed by leaders
- Talented people enabled to do their best - a critical success factor
- Innovation must go beyond today and look to New Markets / Creating New Businesses

Principles / Philosophy – My Top 10

- Strive for balance
- Pursue multiple avenues to maximize value from R&D investment
- Know "who you are" and what your "recipe" is
- Select early stage projects based on broad criteria together with a large dose of experience
- Mistakes happen – don't try to eliminate them

Principles / Philosophy – My Top 10

- Great leaps are very risky
- Effective change is evolutionary …requiring continuous evaluation and feedback processes
- Business silos achieve focus and intensity - their long term preservation leads to a steady decline in technology enrichment, breadth and connection
- R&D must be aligned with business strategy but have enough independence to balance today's needs with the opportunities of the future
- Don't forget the social aspects – people have to work well together

Other Bits of Innovation Wisdom - I

- The number of ideas for a robust pipeline >>> number than most believe are necessary
- Idea generation must be global with multiple sources and connections
- An unstaffed bench of ideas and opportunities promotes greater realism in assessment of current projects
- Multi-disciplinary teams of competent, motivated people beat individual genius every time … more productive, moderate risk, and compress cycles
- Sufficient rate of new business creation requires both linear and non-linear innovation pipeline flow
- Technology, marketing, manufacturing in it from the start

You get the business where you got the business

Other Bits of Innovation Wisdom - II

- Balance the deep scientific competencies with deep engineering competencies – both are needed for success
  - Great outcomes from meshing the engineering mindset with the scientific mindset
- Inject advanced engineering into the early stages of the innovation process to create and adapt more effective process invention
  - Process invention has been equally important as material invention for Corning's innovations
  - Corning embedded the advanced engineering group in the R&D center and hired PhD engineers in multiple engineering disciplines
- Reduce the learning cycles on process engineering with pilot scale laboratory operations
  - Goal: Every major manufacturing process platform in the R&D lab

Other Bits of Innovation Wisdom - III

- Modeling and simulation … and materials and process characterization are critical enablers for materials and process engineering success
  - Yield better fundamental understanding at all levels of the innovation process, create strong IP positions, and often lead to significant cost reduction
- Faster clockspeed innovation opportunities require more technology presence at the customer interface
  - Reduce the learning cycles and requirements iterations
  - Application development, applications engineering and product/process engineering are prominent roles at the customer interface
- Gross margin (GM) performance is really the name of the game
  - GM% is what actually funds RD&E activities
  - Process engineering innovation plays a major role in improving GM% and extending profitable business life cycles

Talent … Talent … Talent

- RD&E people are different than the prevailing business culture
- Continuous infusion of new talent is refreshes any organization, especially R&D
- Development of well balanced innovation leadership demands multiple experiences as early as possible
- Teams prevail
- Different talent is needed at different stages
- T-shaped people with depth and breadth

Successful Innovation
The Critical Role of Leadership

- Capabilities, processes, culture, and money are critical
- But it takes *individual will* to lead innovation
  - **determination and resolve in the face of uncertainty and change**
  - **acceptance of failure and comfort with scientific risk**
- *And individual skill*
  - **the backgrounds of our top leaders must reflect**
    - **real skills in technology and science**
    - **understanding – and rich intuition -- of how and why technology creates business value**

CORNING | Science & Technology Corning Restricted | 43

At the End of the Day..........

- Innovation is Easy to say, Hard to do
  - It is all about the "how": culture, leadership, talent, and process … and patience … and money
- A balanced innovation investment
  - Grows the existing businesses
  - Creates new businesses
- Still plenty of opportunity for invention and innovation in glass, ceramics and other inorganic materials
  - Especially when integrated with other technologies

CORNING | Science & Technology Corning Restricted | 44

CORNING

(137a) The Future of Fuels and Alternative Feedstocks - Recognizing Hype vs. Practical Limitations

William Banholzer, Dow Chemical Company, Midland, MI

The world now agrees that the ideal future requires 100% sustainable sources of energy and product feedstocks in adequate amounts to support a high standard of living for all.

Which new pathways and technologies will emerge to transform our situation? This question is addressed from the perspective of the chemical industry, which was built on oil, natural gas, and coal. These have served as the major raw material feedstocks and energy sources for driving reactions and separations. The industry is now shaping its transformation to sustainable energy and is developing new materials and solutions for energy supply and conversion.

Here we consider the mass and energy balances, capital investment and resource requirements of several key alternative energy and feedstock technologies. These considerations determine where we can expect realistic progress toward sustainable chemistry in both the short and long term, and where we should place our investments.

See more of this Session: Annual Meeting Plenary - Chemical Science Innovation: Future of the US Chemical Enterprise II (co-sponsored by the Council for Chemical Research)
See more of this Group/Topical: Liaison Functions

(137b) Importance of Human Capital: Why Major in Chemical Engineering?

John Anderson, Illinois Institute of Technology, Chicago, IL

Innovation is driven by people who are smart, creative and assertive. One strategy for innovation in the chemical enterprise is to attract people like this to chemical engineering and chemical sciences. Opportunities for careers in technical fields are much different for today's students than five decades ago when I chose a major. Compounding the challenge is that engineering in general has lost market share of students at universities. This presentation attempts to stimulate thinking about steps that higher education and industry can take to attract future innovators to the chemical enterprise and meet their educational objectives.

See more of this Session: Annual Meeting Plenary - Chemical Science Innovation: Future of the US Chemical Enterprise II (co-sponsored by the Council for Chemical Research)
See more of this Group/Topical: Liaison Functions

(137c) The Future for Emerging Energy Technologies

Anthony Cugini, U.S. Department of Energy, National Energy Technology Laboratory, Pittsburgh, PA

The Department of Energy's Office of Fossil Energy, through the National Energy Technology Laboratory, supports a variety of R&D initiatives to develop alternative clean fuels and chemicals from fossil sources, including coal, natural gas and biomass. A key R&D element of the current Fuels Program is the development and demonstration of advanced separation membranes for the production of hydrogen in advanced IGCC power systems for subsequent use as a turbine fuel or conversion to environmentally benign fuels and chemicals. Other research areas focus on the development of modifications necessary for combined coal/biomass conversion which include co-feeding to the gasifier, evaluation of the differences in reaction kinetics and reaction products of the gasification process and the effects of additional biomass contaminants on downstream water gas shift (WGS) and Fischer-Tropsch (FT) conversion catalysts. This presentation will provide an overview of DOE's ongoing R&D initiatives for the co-production of power and fuels/chemicals from coal and coal/biomass mixtures in advanced IGCC systems.

See more of this Session: Annual Meeting Plenary - Chemical Science Innovation: Future of the US Chemical Enterprise II (co-sponsored by the Council for Chemical Research)
See more of this Group/Topical: Liaison Functions

(137d) Chemical Product Life Cycle Management and Enterprise Environmental Stewardship

Monty Alger, Air Products and Chemicals, Inc., Allentown, PA

This presentation will analyze how sustainability affects the chemical industry. The industry has moved from compliance with safety, environmental, and quality requirements to proactive stewardship of products throughout their life cycles. Sustainability -- the integration of environmental, social, and economic considerations -- has become an important element of business management and product development strategy.

See more of this Session: Annual Meeting Plenary - Chemical Science Innovation: Future of the US Chemical Enterprise II (co-sponsored by the Council for Chemical Research)
See more of this Group/Topical: Liaison Functions

(205a) Rapid Development of Coal Chemical Industry in China Providing a Rewarding Career for a Chemical Engineer

Zhenping Lu, Shenhua Coal to Liquid Company, Shenhua Group, Shenhua, China

Abstract currently unavailable.

See more of this Session: CACS 30th Anniversary Symposium: Global Career Opportunities for Chemical Engineers
See more of this Group/Topical: Management Division

(205b) Global Career Opportunities for Chemical Engineers

Kelly Seibert, UOP LLC, Des Plaines, IL

Abstract currently unavailable

See more of this Session: CACS 30th Anniversary Symposium: Global Career Opportunities for Chemical Engineers
See more of this Group/Topical: Management Division

(205c) Career Opportunities for Chemical Engineers in Singapore

Jim Yang Lee, Chemical Engineering, National University of Singapore, Singapore, Singapore

Abstract currently not available.

See more of this Session: CACS 30th Anniversary Symposium: Global Career Opportunities for Chemical Engineers
See more of this Group/Topical: Management Division

(205d) Career Opportunities for Chemical Engineers in India

K. Venkataramanan, Larsen & Toubro Limited, Mumbai, India

abstract currently unavailable

See more of this Session: CACS 30th Anniversary Symposium: Global Career Opportunities for Chemical Engineers
See more of this Group/Topical: Management Division

R&D Management from Beginning to End

Mukund Karanjikar
Senior Associate
Technology Holding LLC
Salt Lake City, Utah

Jack Hipple
Principal
Innovation-TRIZ
Tampa, FL

Eldon Larsen
Professor of Engineering
Marshall University
South Charleston, WV

Extended Abstract

This workshop focuses upon managing R&D projects from beginning to end. Beginning refers to mental picture of an idea whereas end refers to solving the customers need and in the process generating profit for the organization. Using a case study, the speakers will delineate challenges involved at every stage of R&D project from a mental image of an idea to a profit making product. The talk also compares old paradigm of innovation management with the new proposed paradigm. The probability of success can be significantly improved using the new paradigm.

Successful Innovation Management

We have seen many idea management flow sheets, gate management processes, and prioritization recommendations starting with an idea that already exists, but no discussion of where the idea came from or what was necessary to generate and frame it. Moreover attention is not paid to shaping and propagation of the idea. This has been especially true over the past decade when processes such as Six Sigma came into vogue whose main thrust was to eliminate deviation. New concepts and innovation are, by their very nature, deviant! In order to generate fresh and innovative ideas we must create a new thought pattern that may challenge the existing status quo in any number of ways. These can even include, at an extreme, replacing one's current products and businesses with totally new approaches and models.

The key elements of successful commercial innovation include:

1. Continuous flow of unmet needs, both from current customers and related to core competencies of the organization ("outside in" as well as "inside out"
2. Using clear mental models of problems and potential solutions
3. Perseverance for idea propagation
4. Clear definition of the business opportunity
5. An appropriate product development and management team
6. Efficient process scale up and continued input from the market
7. Successful project management to insure implementation on schedule and within budget

We will look at each one and discuss the important issues.

1. Continuous Flow of Unmet Needs and Exploration of Uses for Core Competencies

An organization needs to have a process for how it decides what to work on. This may sound simplistic, when this is left to intuition or accident; it is unlikely that the "best" projects will be chosen.

Steps to make sure that an innovation effort is capturing the best opportunities:

- Talk to groups that are not your current customers but that might be able to use the functional capability that you have as a core competence. Customer competitors are another valuable source. The customers of your customers are another excellent source. This insures that you have awareness several levels above your product and customers and alerts you to potential long term needs and trends that may not be obvious to your direct customer or may not be readily disclosed by them.

Figure 1: Level Diagram

In these discussions we want to make sure that we have a clear definition of what the "ideal result" that our customers are desiring. We can also look at parallel universes of technology and markets to explore opportunities.

2. Using Clear Mental Models of Problems and Solutions

It is one thing to "invent" a new product, but another to make sure that we understand how a potential customer is going to use the product as well as how the customer's customer is going to use their product. The days are gone when we can just make something, put it in a drum or tank car, and wait for the invoice to be paid. We must be intimately familiar with the product's use, its limitations (including regulatory issues if appropriate). We must understand the details of the physical use of the product and how it affects a customer's products and business.

It is also important to make sure that we do not have blinders on when thinking about use of our product. It is rare that there is only one way to accomplish a function. We must be constantly vigilant about competitive activities, not just from direct competitors, but from parallel approaches to meeting a customer's needs.

3. Perseverance for Idea Propagation

New business and product development is hard work and it is easy to get discouraged. An idea that starts our with a 10% probability of success can raise its odds to over 75% by the proper, early stage identification of what challenges must be overcome. If these are not clearly defined, then a new product or business development will flounder and struggle. The perseverance battle rests more on the shoulders of individuals than the organization or top management. Idea has to propagate and the best person to do so is the idea owner. Ideas may be facilitated top-down (by management) however propagation is always bottom up (by employees). A proprietary tool (YouSNAP®) is being proposed to better manage idea propagation. The five elements of the tool are:

a. You – Know yourself
b. Surroundings – Know your surroundings
c. Network – Develop your network
d. Approach – Decide an approach
e. Persistence – Be persistent and iterate

Employees need to understand a simple truth. Ideas are dime a dozen and never bought on face value. Frustration is part of innovation. One should know that failures will exist. Hence know yourself. Are you up to it? Then you should know your surroundings. Often times, a tendency exists in corporations to blame the corporation itself. What is corporation? It is made up of people like yourself trying to serve the customers. If there are non-cooperative people around you, the journey would be commensurately difficult. However the important thing is that you should know this before. This leads to having a robust network of supporters. Once you have done the necessary preparation, decide upon an approach. Finally be persistent. The presentation will discuss the tool in more interactive format.

4. The Business Opportunity

A new product is not a business unless we have a way of sustaining the product's sales, expanding its utility, and developing and enforcing an intellectual property portfolio. As we commercialize a new product, we must be constantly aware of, and be willing to address with resources and commitment, issues such as change in product form, packaging and shipping issues, product purity, and shelf life/stability. These issues are most adequately addressed by constant and continuous communication with the customer.

5. Product Development Teams

We frequently choose product team members based on "who's available". While this may be expedient, it may not serve the best interests of the new business development. The obvious component of team selection is function—marketing R&D, manufacturing, legal, sales, etc. but we frequently ignore the human side of team functioning. If team members spend too much time dealing with personal issues relating to other team members, that energy is not applied to successful new product launch. Tools that can assist in this include measurements such as the Myers Briggs family of instruments, the Hermann Brain Dominance instrument (HBDI), and the Kirton KAI assessment. Knowledge of these personality profiles will not change natural behavior, but can make team members aware of others' preferred styles and open discussion can eliminate hidden barriers to effective team functioning.

6. Efficient process scale up and continued input from the market

Often times, companies strive for launching the best product. It is important to realize that continued input from the market is critical in order to have that 'best product' in the market. It is necessary to efficiently scale up and launch a product sooner than later. It should be perfected using market feedback. Getting market feedback is the only way to close the R&D loop and serve your customers to the fullest.

7. Successful Project Implementation and Management

New product development does not end when the first semi-commercial sample has been OK'd by the customer. A new type of team is now required to oversee the details process design, plant construction (including all the sub issues of permitting, EPA regulations, etc.). Quick and efficient plant start up with appropriate contingency planning is also a part of this effort. Contingency planning, not only related to unforeseen plant construction and start up difficulties, but also to taking care of the customer's needs in these situations must be part of the plan.

Three keys to successful project management lead to successful project completion. These keys involve

- The importance of people
- The importance of defining and setting good objectives
- The importance of good planning

We will apply these three keys to successful project management. This is all about application, not just theory. Today, you can apply the lessons you will learn.

A case study will be discussed in the presentation.

(288a) Managing Your Career

Wendy Young, GE Water & Process Technologies, El Lago, TX

Powerpoint to be added later

See more of this Session: Developing Leadership Skills
See more of this Group/Topical: Management Division

Paper Unavailable

Who Are You Leading? How Do You Know?

Jack Hipple

Innovation-TRIZ
Tampa, FL

Successful leadership requires a clear vision, committed followers, and resources. Without all three the leadership goal may not be accomplished. Very frequently, a clear vision is established and resources are provided but the followers do not follow due to a disconnect between the style of the leader and the styles of the followers and team members. This situation can be resolved by changing the followers, but this can be costly and time consuming. It is not always possible for a leader to choose or change the followers that are needed. Command and control is one approach and the success of this can be a function of the supply and demand of the job market at the time. A far more reliable and robust approach is for the leader to take the time to understand his or her own style and behavior preference as well as that of their followers. Understanding these can help a leader be successful under any circumstance with any group. This presentation will overview some key behavior preference and style measurements that accurately assess social and problem solving styles and how they can be used to assist leaders and followers in achieving this goal.

Leadership and People

We use this term all the time without giving sufficient thought to what it really means. It may mean getting a group of people to see something that is not obvious to them, but is important to an organization and its senior management. It may be responsibility for leading a multi-national and multi-cultural team in building a multi-billion dollar capital project involving thousands of engineers in many locations. It may mean trying to figure out what the next new business, product, or service an organization should invest in when there is no clear indication of what it should be. However, all these responsibilities, and many other leadership roles involve one thing in common—the influencing and direction of people. A leader frequently does not get to choose who is on their team and occasionally needs the support of people who do not directly report to the leader. People are different. No two individuals are the same. They have technical backgrounds, different professional experiences, different personal preferences, and different biases. In order to lead a small team or a large multi-national group, it is imperative that a leader understand these differences for two primary reasons:

1. Task assignment. Everyone is not good at everything. Everyone has different preferences for what they would prefer to do. Assigning people responsibilities to people who do not enjoy that type of work when not necessary will only produce a drop in productivity and team morale.
2. Team relations. The time that a team spends together is valuable and expensive. If people on a team and its leader do not understand the differences in styles and

preferences, a certain percentage of the time and energy naturally present within each individual and the team will be wasted in arguments and debates relating to perspective, objectives, resources, etc. There will never be an answer that everyone will agree to, but an understanding of why a particular individual feels the way they do can go a long way to making everyone feel that their contribution and perspective is appreciated.

Team Member Preferences and Styles

Over the years, many different individual assessment instruments have been developed which are useful for assessing differences in social and problem solving styles. Both are important in a typical engineering team as the leader would like the team to "get along" but also solve a problem or design a process, product, or facility. These are different things. Social style relates to how individuals relate to each other in a social context, not necessarily one that includes technical problems. Problem solving style relates to how individuals analyze problems and how they solve them, independent of who they may be working with who is also involved in the problem solving or design challenge. It is possible for individuals to have identical social styles as measured by an assessment such as the Myers Briggs assessment and to have very different problem solving styles as measured by an assessment instrument such as the Kirton KAI™

Social Styles

The Myers Briggs MBTI™ was the original social style assessment instrument and has been given to millions of individuals. It has successors such as Insights™ and 16Types™ which provide feedback in the same summary form. Regardless of which instrument is used, a person's social style can be characterized along the following continuum:

Extroversion/Introversion (E vs. I) This describes the outward vs. inward focus of an individual. An extrovert enjoys socialization, verbal communication, cold calling, possibly parties, and other occasions and events where outward communication is enjoyed and expected. A strong introvert, on the other hand, may prefer to read a book, surf the web for information and become very nervous in social situations where direct oral communication seems to be required.

Sensor/iNtuitor (S vs. "N").Note: An "N" is used here, rather than an "I" to avoid confusion with the "I" used in the previous category. A Sensor is strongly concerned with data, facts, and measurable aspects of a problem or situation whereas an Intuitor is equally concerned with aspects of a situation that may not be directly measurable such as implications, indirect effects, gut feelings, emotions, and undefined possibilities that cannot be immediately quantified. This characteristic primarily applies to how information and data is taken into a mental analysis as opposed to what comes out afterwards. A simple question to ask someone to find out which one of these they may be is to ask them to describe a leaf. A sensor will say things like "green" or "scaly", terms that are physically descriptive. An intuitor will describe it as "light" or "reflective".

Thinking/Feeling (T vs. F) This attribute characterizes how someone treats the analysis of the information received. A "T" person will be much more concerned with the impact of a decision on economics, structure, and costs while an "F" person will be just as strongly concerned about changes might affect the personal lives of individuals and their careers. In a downsizing situation, the "T" will want to know how much money will be saved and what the outplacement services will cost, while the "F" will want to know about the impact on peoples' families and their psychological well being.

Judging/Perceiving (J vs. P) This characteristic describes the focus of an individual and their approach to implementation and planning. A "J" will be very concerned with schedules, commitments, deadlines, and critical path diagrams. A "P" will be much more open to loose schedules, leaving the door open for changes and more comfortable with uncertainty. Deadlines are critical to the "J" and far less important to the "P" who is more worried about missing possibilities that cannot be seen as opposed to meeting commitments to do what is known.

You can imagine, as a leader, the challenges in managing a team where these characteristics were not uniform. It's easy to work with and manage people that are like us, but that does not necessarily make for the best team structure. If everyone thinks alike, there is no one to ask the difficult questions or propose alternatives and options

The social style assessment will come back, independent of the type of instrument used, as four letters, such as ESTJ or INTP. Though the descriptions above emphasize the extremes, it is possible for an individual to be borderline or in the middle. There can be small changes in an individual's preference over time, but not large ones. This type of assessment also provides an insight as to how a person will respond under stressful situations. For example, a strong introvert will become very extroverted when fundamental beliefs or values are questioned. Obviously, this type of situation can occur within a project team and if social style preferences are known and understood, these types of situations can be handled more effectively.

Problem Solving Style

An individual's problem solving style can be measured very accurately with an assessment instrument known as the Kirton KAI™. This assessment, consisting of 32 simple questions, analyzes a person's approach to problem solving. The assessment comes back as a number from 32-160 with an average of around 90 and a 2 sigma deviation between 70-110. A lower KAI™ number would describe someone who problem solves with known parameters, likes schedules and agendas, and generates "filtered" ideas whose probability of being practical is high. They are described as an "adapter". At the other extreme would be someone who tends to generate a larger quantity of ideas without necessarily worrying about their practicality, is less concerned about agendas and schedules, and will "bi-associate" ideas and make connections between disparate concepts not made by others. In KAI™ jargon, this person is described as an innovator. A simple question to ask someone to get a sense of where they are on the KAI™ scale is what they would do as a tea bag manufacturing plant manager who has just lost a major customer amounting to over 2/3 of the plant's business. An adaptor will suggest ideas relating to cost reduction and layoffs and encouraging sales and marketing to find some more tea bad customers. An "innovator" will ask what new product ideas might come from packaging something else in the bad or what else the bag fabric could be used for.

As mentioned earlier, people with the same social style profile can have differing problem solving style profiles. This can set up a situation where a team leader thinks that everyone on a team "likes" each other and their job will be easy, but if the team's challenges include serious problem solving, things may not be as easy as anticipated.

Leadership Situations

Let's consider some situations and the implications of not understanding the nature of the team being "led". First, an international new product introduction. If the leader's future income is dependent upon how fast and how many countries a new product can be introduced into, there is a high level of risk, especially if manufacturing and distribution are under contract in some sense. If the team members do not have the same stomach for risk and "fuzzy" situations, they may throw up artificial roadblocks and record "remember I told you" notes to be vindicated later if the introduction does not go as planned. If the leader knows about this style preference ahead of time, then a rational discussion can be had regarding how the risks can be minimized.

Consider a multi-country engineering team whose meetings may be as often over the web as in person. These various social and problem solving types are not uniformly distributed and can be affected by cultural biases. It is critical, especially when deadlines and commitments are involved, to make sure that a leader ensures that the meaning of deadlines is clear. A "J" will understand easily while a "P" may interpret a deadline as simply a target or a goal. A high scoring KAI person may take risks that seem acceptable to him/her but may not be acceptable to the team leader. Interpretation of boundaries will vary and the language needs to be understood at the beginning of a project, not in the middle of a crisis.

To summarize, it is critical that a team leader understand both social and problem solving styles of the people on their teams. Without such an understanding, there is great risk of misunderstood goals, decision making based on misunderstood criteria, and problem solving done with either too little or too large a boundary condition.

(288c) Strategies for Motivating Employees

Charlene A. Wall-Warren, BASF, Florham Park, NM

Strategies for motivating employees

See more of this Session: Developing Leadership Skills
See more of this Group/Topical: Management Division

Paper Unavailable

(288d) Successful Networking-Yes, Engineers Can Network Too!

Lori McDowell, Praxair, Inc, The Woodlands, TX

The author will discuss networking.

See more of this Session: Developing Leadership Skills
See more of this Group/Topical: Management Division

Paper Unavailable

(288e) Dealing with Uncertainty In Decision Making

Diana Matonis, Flucomp Enterprises, Hickory Hills, IL

What is the difference between risk and uncertainty?

What are some quantitative methods that aid decision making under uncertainty?

What are some qualitative methods that aid decision making under uncertainty?

How can decision makers effectively combine qualitative and quantitative methods?

See more of this Session: Developing Leadership Skills
See more of this Group/Topical: Management Division

Paper Unavailable

(288f) Effective Communications

Gautam Pillay, Department of Chemical Engineering, Rowan University, Glassboro, NJ

Effective communication between management, employees, external stakeholders, and partner organizations (often geographically separated).

See more of this Session: Developing Leadership Skills
See more of this Group/Topical: Management Division

Paper Unavailable

(288g) Speaking Truth to Power

Deborah Grubbe, BP Products North America, Inc., Chadds Ford, PA

Speaking truth to power

See more of this Session: Developing Leadership Skills
See more of this Group/Topical: Management Division

Paper Unavailable

(288h) Dhamma Tools for Managers

Gavin Towler, UOP, Des Plaines, IL

This paper will describe methods for maintaining your composure, motivating your team, dealing with obstacles, getting results and becoming a better leader, while maintaining a harmonious working environment.

See more of this Session: Developing Leadership Skills
See more of this Group/Topical: Management Division

(458a) Historical Perspective On Consumer Goods

Tasha Desai, Henkel Consumer Goods, Scottsdale, AZ

In the consumer products industry, convenience for the consumer motivates a lot of innovation. Over the years, detergents have become multifunctional, stain removers have become sprays or foams, and packaging for household cleaners has evolved to provide more convenience to the consumer. Historically, it is apparent how products have slowly changed as consumers began to prefer easier to use and multi-purpose goods.

See more of this Session: Innovation In Consumer Product Management and Design
See more of this Group/Topical: Management Division

Using Other Peoples' Brains to Invent New Products

Jack Hipple
Principal
Innovation-TRIZ
Tampa, FL

Extended Abstract

Inventing and commercialization of new products is a risky proposition. One of the ways of minimizing costs and time to market is to use technology approaches from parallel universes of technologies that are non-competitive in a commercial sense. "TRIZ" (Theory of Inventive Problem Solving) is a methodical way of identifying parallel functional universes where the same result and function is used, but possibly for a totally different commercial reason.

What is the Function?

This seems like such a simple question to answer and it may be. However, we frequently answer it in highly specific technical terms unique to the industry or product area of concern. Let's look at an example. In the air traffic control business, FAA regulations are requiring more and more information to be displayed on air traffic control displays. This is done without thinking about the possible consequences of an air traffic controller trying to make quick decisions based on graphical display information which may cause information overload. If we were looking at a parallel universe which has a very similar problem, where would we look?

In a real problem solving situation with an air traffic control display manufacturer, the following technology areas were not even considered or reviewed in attempting to come up with new concepts relating to the air traffic control display challenge:

1. Chemical process control displays
2. Hospital operating room displays
3. Video games

If we frame the challenge as one of increasing the ability of an air traffic controller to respond to information change or emergency situations, our view of possible ideas is rather limited. For example, if we did a web search with the heading of "air traffic control displays, we would find about 4.7 million hits. However, if we thought for a minute about the function being performed, we might type in something like, "video displays", we would find almost a billion hits! We would industries such as chemical process control displays, heart surgeon operating room displays, video games, and many other products that have exactly the same problem.

Let's take another example in the auto industry. Though head rests in cars are mandatory, they are seldom neither comfortable nor easy to adjust. In a Wall Street Journal article highlighting the efforts of the Ford Motor Co. to address this problem, it was shown that the physics of baseball catchers' mitts have exactly the same problem. If we consider our web search again and type in the word "human headrests", we will find about 150,000 hits whereas if we type in "stop moving object", we will obtain 14,000,000 hits—nearly two orders of magnitude more. This article highlighted the fact that the physics of catchers' mitts (used since the late 1890's) had the same physics involved in their design and had never been considered a model for design of headrests.

So the first lesson when looking for parallel universes of knowledge is to use the most general description of the function being performed. Using language that a ten year old would understand is a good general guideline.

TRIZ Thinking and Analysis

"TRIZ" is a Russian acronym standing for "Theory of Inventive Problem Solving" and is an algorithm based problem solving technique based on the fact that when we study the global patent literature, we can only find 40 inventive principles that can be used to describe any patent, regardless of its field. Thinking about this fact in reverse means that the solution to most problems already exists. The exception to this would be the area of pure science such as new chemical or biological entities. It is very difficult for problem owners to want to admit that their problem is not special and unique. However, once this psychological barrier has been overcome, and the problem generalized, then it becomes easy to identify parallel universes of technology and markets where a problem has already been solved

Application in Consumer Products

Let's consider some applications in the consumer products area.

In the design of catsup bottles, Heinz finally discovered the force of gravity and redesigned their bottles to, in effect, flow upside down so that their high viscosity catsup formulation would more easily flow. Several years later, shampoo manufacturers inverted their bottle designs to take advantage of a free resource. Why was there a time delay?

In the paper towel industry, it was observed that many consumers do not need a normal sized paper towel for small spills and a second layer of perforations were incorporated into Bounty paper towels to allow the tearing of a towel into a smaller piece if desired. Sometime later, cheese manufacturers such as Kraft discovered that perforating cheese slices would allow consumers to tear off cheese squares in just the right size to fit on a cracker.

In another interesting parallel universe example, the same basic technology used in the biodegradable trash bag industry is now being used in dissolving heart stents under development in the pharmaceutical area. This did not happen until several years after the trash bags were on the market and their patents issues. There are two interesting questions here. Why did the trash bag formulators think about higher value added markets for their formulations? Second, why did the heart stent manufacturers not look in the garbage area for technology (possibly too low tech?). If we go through the same web search exercise, we would find the following results:

Disappearing stents	12,000 hits
Dissolving plastics,	1,000,000 hits
Disappearing plastics	2,200,000 hits

The point in showing both "dissolving" and "disappearing" is that dissolving is only one way of making a material disappear! What might be some others?

Summary and Conclusion

When analyzing a new consumer product opportunity or a new product development opportunity more generally, it is critical to analyze the product or need from a functional, not a product, standpoint. This will allow a more general consideration of other approaches and technologies that may already exist to solve the problem or design the product.

(458c) Facing the Top Five Innovation Barriers for Consumer Products: Systems and Strategies

Jeff Lindsay, Asia Pulp and Paper, China, Shanghai, China and Cheryl Perkins, Innovation Edge, Neenah, WI

In our experience with consumer product innovation, several common innovation fatigue factors repeatedly pose special challenges that can interfere with innovation success, especially in large companies. We explore these challenges and offer solutions that can strengthen culture, identify and eliminate bad behaviors that defeat innovation, and strengthen judgment in pursuing cost-effective innovation. Principles that must be understood by managers as well as researchers and product developers will be reviewed, with case studies drawn from global corporations as well as start-ups.

See more of this Session: Innovation In Consumer Product Management and Design
See more of this Group/Topical: Management Division

(458d) The Science Behind the Innovative Laundry Sheet Purex Complete 3-In-1

Matthew Petkus, Laundry Care R&D, Henkel, Scottsdale, AZ

Consumers are king when it comes to the consumer goods industry. This is why products that are engineered for convenience take precedence in the category. In the fabric care category there was a need to develop a product that delivers cleaning, softening, and antistatic in an all in one laundry product. That is why Henkel designed and launched a novel laundry sheet that offers all of these benefits. The science and design behind the Purex Complete 3-in-1 laundry sheet was imperative to its successful launch.

See more of this Session: Innovation In Consumer Product Management and Design
See more of this Group/Topical: Management Division

(458e) Rational Design of Pharmaceutical and Other Liquid Formulations

Elisa Conte[1], **Sascha Sansonetti**[1], Peter A. Crafts[2] and Rafiqul Gani[1], (1)CAPEC, Department of Chemical and Biochemical Engineering, Denmark Technical University (DTU), Kgs. Lyngby, Denmark, (2)Process R&D, AstraZeneca Pharmaceuticals Ltd., Cheshire, United Kingdom

Every year pharmaceutical, health-care and related industries invest a high percentage of their total time and resources on development of new products and improvement of existing (pharmaceutical) formulations. Product development is still mainly based on trial and error experimentation and laboratory procedures, indicating long research processes without any certainty of a definitive solution and far from achieving the *optimal formulation* of the product. Besides, (pharmaceutical) ingredients to be used for the experimentations are often very expensive and not available in the desired form or at the desired level of purity. Moreover, the experimental procedures are affected by unavoidable errors due to the complicated and often not-standardized procedures. Therefore, the need for a rational and systematic model-based approach to plan suitable experimental designs would constitute a fairly useful tool for pharmaceutical product design and product design in general. A model-based approach significantly reduces the number of experiments necessary for product design and at the same time gives a clear idea about the direction to take for further (reduced) experimentation. In such a way tedious and uncertain screening processes can be avoided with sensible reduction of both time and resources for the product launch on the market.

Identification of a pure solvent or anti-solvent for a specific Active (Pharmaceutical) Ingredient (API or AI) is a problem of major concern for the pharmaceutical and related R&D departments. Solvents, lipids and other compounds are commonly employed in product formulations as well as in API processing. In addition, the screening of several solvent mixtures that sometimes show improved characteristics of solubility toward a particular API is necessary. Besides, in liquid formulated products, API (or AI) needs to be mixed with additives to enhance the product quality and solvents to deliver the product.

This work presents a model-based method for solvent screening and solvent mixture design for pharmaceutical applications and API/AI based liquid formulated products. Two cases are investigated: a case where the solvent or solvent blend is designed for API processing (here the solvent is not part of the final API based product), and a case where the solvent or solvent blend is part of the API/AI based formulated product. In both cases, the solubility of the API/AI needs to be predicted in solvents (single solvent or blends). A general solubility modelling procedure having different options for the solubility calculation (such as UNIFAC, UNI-sac and PC-SAFT) is employed. A unique feature of this solubility calculation procedure is that it is truly predictive and does not need any new experimental data for the computer-aided design stage of the formulation design methodology (Conte *et al.* 2011).Once the domain search has been reduced to a few solvents or solvent mixtures, solubility can be computed by correlative models such as NRTL-SAC andPC-SAFT.

The newly developed method described above is to be integrated in a comprehensive framework for formulation design that was previously developed and tested for personal care products. Such a method was implemented in the software 'the virtual product-process design laboratory' (vPPD-lab) (Conte *et al.*, 2011).

Examples illustrating the application of the procedure for solvent, solvent-blend and/or AI-based formulation design will be presented. Also, a conceptual case study validating the models and the design steps involving the design of a solvent mixture that maximizes the API solubility as well as the identification of an optimal anti-solvent for a well-known API such as paracetamol will be presented.

References

Conte E, Gani R, Ng K M, Design of formulated products: A systematic methodology, AIChE J, 2011 (DOI 10.1002/aic.12458)

Conte E, Gani R, Cheng Y S, Ng K, M, Design of formulated products: Experimental component, AIChE J, 2011 (DOI 10.1002/aic.12582)

See more of this Session: Innovation In Consumer Product Management and Design
See more of this Group/Topical: Management Division

(512a) Academics and Open Innovation

Richard L. Long, New Mexico State University, Las Cruces, NM

Open Innovation generally refers to property rights issues associated with orgainzations such as nine sigma. In a broader sense, it refers to the issue of "credit" of which cash awards are one type. The issue of "control" and the financial and other rewards have to do with the relationship between buyer and seller. Transactions must have a degree of confidence in the integrity of the parties and establishing that confidence is a key requirement.

See more of this Session: Challenges In Open Innovation and Growth
See more of this Group/Topical: Management Division

Effective Quantitative Decision-Making based on Lowering Cost of Capital and ROI thru Safety Eng Studies

Gutiérrez Capriles Glenda J
***Risk Project Analysis Leader Engineer Advisor**
Avenida Circunvalación del Sol, Urbanización Santa Paula, Torre Inelectra, piso 7, Municipio Baruta, Estado Miranda, Venezuela
glenda.gutierrez@inelectra.com

Prepared for Presentation at
AIChE 2011 Annual Meeting
Minneapolis, MN, United States
October 16-21, 2011

UNPUBLISHED

Effective Quantitative Decision-Making based on Lowering Cost of Capital and ROI thru Safety Eng Studies

Gutiérrez Capriles Glenda J
***Risk Project Analysis Leader Engineer Advisor**
Avenida Circunvalación del Sol, Urbanización Santa Paula, Torre Inelectra, piso 7, Municipio Baruta, Estado Miranda, Venezuela
glenda.gutierrez@inelectra.com

Keywords: Management of projects, processes of management, areas of management, phases or service life of the project, Decision-making methodology, instrumented protective systems (IPS), spills, return on investment (ROI), hazardous events, fire zones

Abstract

The intention of this article is to reflect on the main aspects that affect the performance of the projects, for such effect took place a diagnosis of the schemes of management used by the company of the consulting sector. For the study one approached the areas of generation business, petroleum, refining, petrochemical, technology, etc of inelectra and the operational area of a production facility.

The process industry has adopted automation to improve product quality and production rates, to reduce the potential for operator error, and to decrease resource requirements. Process automation includes many different instrumented systems, such as process controls, alarms, instrumented protective systems (IPS), and emergency shutdown systems. These systems range from simple hardwired systems to complex programmable electronic systems.

Good design leads to safe operation by anticipating potential errors and failures and implementing means to detect their presence, so that they can be corrected once implemented, the increased complexity may require a significantly greater level of attention during assessment, design, inspection, testing, and maintenance.. Good design ensures that the core attributes of independence, functionality, integrity, auditability, reliability, access security, and management of change are addressed consistent with owner/operator requirements. Poorly implemented design can lead to significant hazardous events, involving impact to people, the environment, and plant assets.

Various valuable information scattered all over CPI (Chemical Process Industry) needs to be integrated and retrieved efficiently to manage technique, facility and personnel. This paper focuses on a new decision-making methodology of safety-related investment and management of main causes for accidents and failures. The methodology integrates safety technology, process data and business information in a petroleum and oil-refining industry. This methodology suggests a quantitative investment procedure to reduce safety accidents and plant failure cases within limited budget efficiently. Major factors are considered as accident history, human error, facility, service, operation, maintenance, emergency plan, and priority of investment, effectiveness of improvement, environmental concerns, loss control and cost.

1. Introduction

Any industry, principally oil-refining industry, have to keep safety management methods that ensure risk of fire, explosion and toxic material release which is very important in saving life, assets and environment. The right decision-making for improving safety in chemical plants requires more quantitative and systematic methods of analyzing previous accident records, process data and business information. This safety engineering studies are necessary to be assure to the whole decision-making procedure for project risk analysis evaluation, to prevent major accidents and to reduce the risk of a plant.

The initial design effort includes the prediction of potential hazards. Numerous types of hazard analysis methods have been proposed and used. Hazard analysis methods are classified by qualitatively and quantitatively. The qualitative methods include Check-lists, What-If Analysis, Failure Modes and Effects Analysis, Preliminary Hazards Analysis or HAZID and HAZOP. The quantitative methods include Event Tree Analysis, Fault Tree Analysis, Quantitative Risk Analysis, Reliability Analysis and Safety Integrity Level (SIL) / Safety Instrumented System (SIS).

New technology is of value only if it provides a suitable financial return. Many articles have been published on the potential economic benefits and return on investment (ROI) of facility and advanced automation technology in the process industries. Unfortunately the benefits claimed are often unrealistic and unsubstantiated; this leads to significant credibility issues when the forecast benefits are not achieved and to a lack of confidence on the part of management on proposals in this area. In this article a short review of the proper way to perform financial analysis as part of project risk analysis that was affected by improper safety engineering studies execution.

2. Review of Plant Economics

The first step is to examine the plant as a financial asset. We may think of plants as a collection of equipment and personnel that convert raw materials into products. From a financial point of view, a plant is an asset that consumes money and produces money, hopefully producing more than is consumed. The major monetary components are shown in the figure following:

Figure 1: The Plant as a Financial Asset

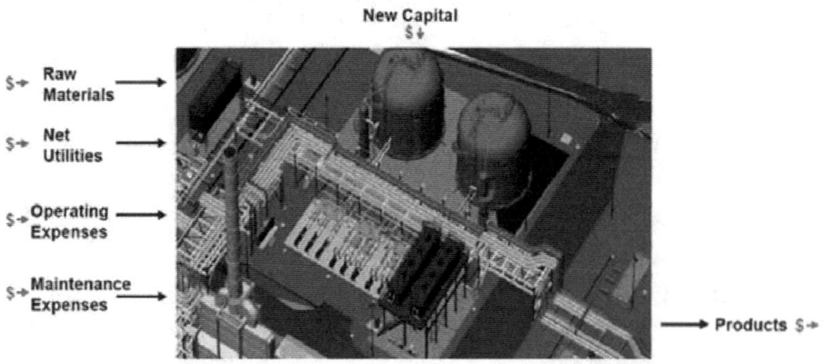

By convention, financial inputs are classified as either capital or expense. Expenses include all the ongoing costs of producing the products such as raw materials, net utilities (used less produced), operating costs, maintenance expenses, and other miscellaneous costs. Capital has two components – working and investment.
Investment capital refers to the cost of major equipment or system additions that will last for several years and can be depreciated for tax purposes. Working capital is the value of inventory and required net short term financial funding.

3. Return on Investment Capital taking into account also, Procurement and Safety

To properly prepare an investment evaluation we must understand the language and objectives of the plant's financial management. In this section a quick overview of the subject is given. For more background and a more comprehensive review.
Ultimately the management of a company has an obligation to increase the long term financial value of the corporation to its owners, who are normally the shareholders. This value can be measured in many ways but stock market valuation is one of the more important and ultimately is a function of financial performance. There are equally many ways of evaluating financial performance but the primary measure used here will be Return on Invested Capital (ROIC). This is not only a good internal measure but also correlates with long term stock market performance. The consulting company McKinsey (2003), recently published results of a study by of 130 publicly traded chemical companies in the US and Europe and their financial and market performance over 40 years. Their conclusion: "only returns on invested capital drive market-to-book valuations".

Figure 2: Return on Invested Capital

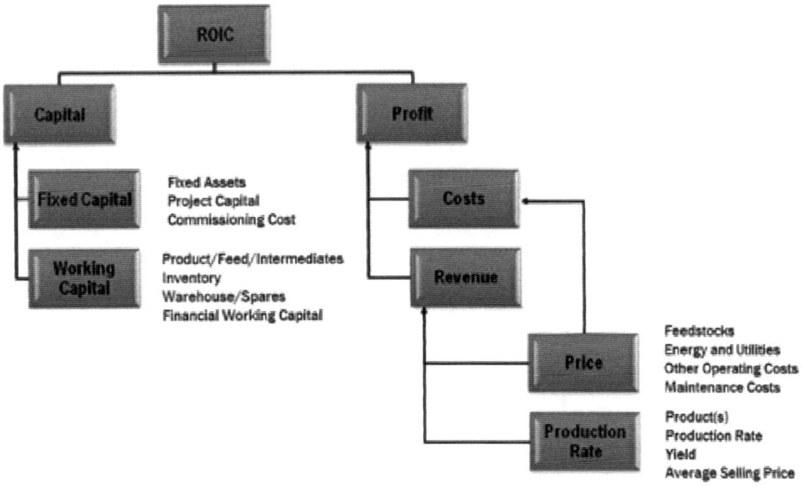

To increase ROIC, capital must be reduced or profit increased or preferably both at the same time. The primary areas where automation and advanced automation savings are normally found are illustrated in the boxes at the bottom of the figure. When a project is considered, all possible savings areas should be evaluated.

Potential capital savings include both fixed and working capital components. Potential project capital cost reductions due to automation choices can include savings in:

- Engineering
- Procurement Costs
- Purchase Price
- Installation, Configuration, Calibration & Commissioning
- Project Execution

Working capital can be reduced by reducing raw material, intermediates, and product inventories and also by reducing owned spares for equipment stocked in the warehouse. Capital deferred also results in savings due to the time value of money.

Deferred capital can result from longer equipment life due to better control or from more production from the same equipment which results in a postponed plant expansion.

Expenses can be reduced by lowering energy /utilities usage and decreasing raw material costs through increasing yields of desirable products. In special cases it may be possible to substitute a lower valued raw material.

There are several types of maintenance costs including:

- Scheduled Maintenance
- Unscheduled Call Out
- Major Shutdown

Reductions can occur in all types of maintenance costs due to improved automation performance and enhanced monitoring of process equipment. Specific savings can include reductions in:

- Unscheduled maintenance
- Number of routine checks
- Time to perform required maintenance tasks
- Maintenance materials purchase
- Number and costs of required tasks during scheduled shutdowns

Reduced off specification material can lower reprocessing costs. Demurrage is the cost charged by ships waiting to load. Waits beyond contracted amounts incur penalties. In selected cases it may also be possible to reduce average staffing levels.

Abnormal event prevention deserves some special comments. Reducing Health, Safety, and Environmental (HS&E) issues is at the top of every plant manger's agenda.

Improved automation is often a key to reducing these types of events. If your proposed investment will result in safer operation or reduced emissions it should definitely be included in the justification description even if no quantitative economic value can be attached to these improvements.

Next is the area of increased average selling price for products. Automation can contribute to this goal by increasing the yield of more valuable products. Reduced amounts of lower valued byproducts will increase average revenue per unit feed. Off specification material, which can be reduced through improved control, usually brings a discounted selling price. Occasionally an

increase in product quality due to improved control will actually permit selling the existing product(s) at a higher price.
With regard to increased production, the key question is whether or not the additional product(s) can be sold. Increases are only financially valuable for production limited plants, i.e. those manufacturing products where the market can absorb the increases. If not, no benefits can be claimed for the improvements. Increased production can result from the capability, with better control, to operate closer to production limits at constant product quality. Increased production can also come from reduced unscheduled downtime for equipment due to better reliability, shorter batch cycle times, lower grade transition time, less product reblending, and decreases in scheduled shutdown duration and frequency. The shutdown frequency can be reduced, for example, by increasing a furnace run time between required decoking or not cleaning a heat exchanger prematurely.
To estimate the potential magnitude of the savings, examine the normal expenditures and look for specific quantifiable areas where savings are possible. Often historical data will provide a basis for this analysis. External consultants with experience in this area can also help insure that potential savings are not overlooked and that there is proper documentation for the estimated savings used for the justification.

3.1. The Cost of Tendering

Major equipment manufacturers have calculated that it costs them up to thousand dollars to bid on a complex tender. These costs are eventually passed on to the customer although be it indirectly.
Among the major contributing factors, are:
Long Lead-Time of the Process - There are many time consuming tasks that contribute to the long lead-time of tendering including: internal collaboration within a multi-disciplinary group of people within a single organization (but often geographically distributed), obtaining approvals (both financial and technical), preparing the tender specifications and documents, interfacing between customers and suppliers, interfacing with other companies, third party consultants, partners in a consortium and sub-contractors, assembling accurate bidding documents and calculating prices and time estimations.
The Quality of Tendering and Bidding Documents - This is a crucial element for successful and efficient tendering and involves criteria such as the clarity of requirements and specifications, compliance with quality certification standards, clarity of the bid, exhaustiveness, adequate structure, nice presentation, compliance with specific requirements set-out in the tender, completeness of the information provided, etc.
Inability to Quickly and Reliably Assess the Bids - Differences and inconsistencies in the bid formats make it very difficult to assess the bids and compare apples with apples. A large amount of time is often spent collating data from the various bidders into a format that allows for proper comparison and evaluation.
Reliability of the Cost and Time Estimate - Ultimately companies who are consistently accurate in their pricing and delivery estimates will win in the long run. The ability to provide reliable estimates depends both upon the capability of the bidders and the clarity and unambiguousness of the tenders. Clear, complete and non-ambiguous tenders will also naturally

lead to lower bidding prices. As uncertainty diminishes so does the "safety price" or padding added to the actual estimate.

Monitoring Delivery of the Works - The ability of the customer to continuously and transparently monitor the current status of the works is a definite advantage in that it reduces delays and ensures delivery in compliance with the contract.

3.2 The ROI of Safety

Although it is difficult to prove that incidents will happen in the future, management and financial representatives of the company need to be aware of the consequences of an ineffective safety solution. By understanding the elements of risk management and the methodology of a safety investment analysis, safety engineers or risk management professionals can present the financial justification to help make operations safer and reduce both workplace injuries and insurance costs.

Integrating a safety solution to manage risk is a process that is best employed via up-front teamwork between operations and risk management, such as risk managers, safety professionals and claim analysts. In addition, it is important to understand risk and the different types of risk that exist on a plant floor. Risk awareness often leads to better decisions that can reduce soft costs. By providing an easy-to-be-safe environment for operators with safety solutions that are designed into the machine and focus on complete machine life-cycle requirements, companies are able to reduce indirect costs. In this paper, we will highlight key elements of effective risk management and the risks that affect the engineering, procurement, construction and operation of Production facility, discuss the implications on workers and provide ideas to measure potential savings that can be found in proactive safety investments.

Risk management is a two-step process that identifies and analyzes uncertainty in decision-making. It consists of determining risks that exist in an equipment investment and managing those risks in a way best-suited to achieve investment objectives. For managers and financial professionals investing in safety, proving the benefit of the safety investment is difficult since it could prevent injury and avoid costs for a company. Using the risk management process, professionals can determine the level of safety investment that meets their needs and would provide the best result for their situation.

An unplanned interruption can threaten a company's supply chain because of the decreasing amounts of raw materials readily available on the plant floor. These limited resources affect productivity levels and operation costs and threaten the overall health of a company as lower levels of inventory put manufacturers at increased risk for losing revenue each time a production line goes down. Effective safety programs can be implemented to help minimize this risk so manufacturers can better protect their supply chains.

Once the risks are fully defined and clearly understood, they must be designed out or mitigated to the greatest extent possible. Risk mitigation measures are necessary in order to physically improve the machine to prevent personal injury or property damage. Risk mitigation can be accomplished through a variety of activities, as demonstrated in Figure 2 below. One effective mean is to use safeguarding equipment such as installing physical guards or safeguarding control products, such as light curtains, safety relays and cable pull switches to reduce risk to employees.

Figure 3 – Hierarchy of Protective Measures

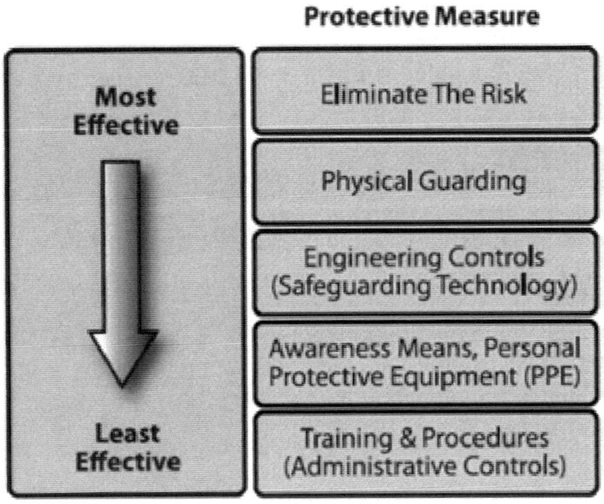

Using a formal risk assessment process also provides the benefit of documenting how the risks have been reduced or designed out via safeguarding solutions or other protective measures. The use and documentation of the risk assessment process is a critical tool for companies to show their due diligence and good engineering practices in providing a safe work environment. By effectively implementing and documenting the process, the resulting protective measures, and training programs, a company can potentially lower risks of litigation should an incident occur. After implementing and documenting the process, it is important to provide the appropriate training and supervision. It is crucial to that operators understand the safety measures, including usage of personal protective equipment. Operators must be trained.

Safety is not a luxury reserved for developed countries. One topic of discussion is this idea that safety and health programs benefit a company's bottom line. For many analysts, the bottom line is simple: "The CEOs 'who get it' recognize that a corporate commitment to safety has its costs, but, more important, it pays tremendous dividends in business and human capital."

So there is a clear connection between safety, quality, production, and profitability."

4. Investment Costs

Next investment costs must be estimated. When performing investment analysis, it is important to consider the full life cycle costs, not simply the initial purchase price, of new equipment and new software systems. Many studies have shown that more than two thirds of their life cycle costs occur after installation.

In the project initial, or capital costs area, include the following:

- Installation costs
- Required infrastructure upgrades
- Software licenses
- Application specific services – design, configuration, installation, integration and commissioning
- Project expenses – project management, purchasing
- Training – System support and end user

Also, ongoing maintenance and support cost estimates. Don't forget the last item. Make sure that you budget for ongoing training. The best long term job security today is maintaining you and your staff's skill levels and expertise current. It is easy to fall into the trap of being so busy with day to day problems that you lose sight of the longer term requirements.

5. How Can ROI be Useful?

Profits happen when a company operates effectively. We can tell that the management team is doing its job well if the company prospers, obtains funding, and rewards the suppliers of its funds.ROI is the principal tool used to evaluate how well (or poorly) management performs.

6. Safety Decision-Making Procedure

A framework for viewing management information systems is essential if an organization is to plan effectively and make sensible allocation of resources to information systems tasks. A framework is needed to provide a more efficient allocation of resource in safety information. This study suggests the procedure, development and information management of safety decision-making methodology in two different information areas as shown in Table 6.1. A framework is suggested for Decision-Making by using model structure and implementation procedure which differs sharply between the structured and no-structured information area.

Table 6.1. Framework of Safety Information Management System

	Safety	Process	Business
Structured	MSDS Regulations Standards Company Policy	P&ID PFD Plot Plan	Department Personal Facility Warehouse Buildings
Non-structured	Accidents Measures Risk Accident History Causes Analysis	Technical Reports Technical Audits Design Execution with Quality MOC History	Cost Strategy Planning Investment Priority Project Risk Analysis Benefit-Cost Analysis

CCPS (Center for Chemical Process Safety, 1998) in AIChE presents effective taxonomy of safety data for reliability analysis as shown in Figure 4.

The basic building blocks for the reliability analysis fall into two main classes. One is the data regarding specific events occurring in a piece of equipment. The other is the data describing the sample space of equipment. Event data in the first class is used as a numerator in the reliability calculation and inventory data in the latter class is used as the denominator.

Classification of safety accident is important while collecting safety information, so we classified it as accident results, direct and indirect causes of accident, primary causes, process status, utilities, departments etc.

Significance and operating variables also affect the business strategy in plant site also of Project Risk Analysis.

Figure 4. Database Taxonomy and Structure

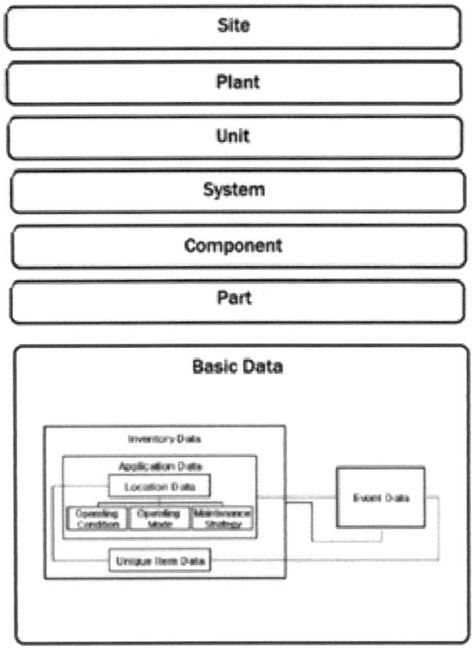

This study describes an overall systematic decision making procedure of preventing or minimizing safety accident under the given budget as shown in Figure 5. To decide the investment of safety related facility, two inputs are required: 1) the priority of investment and 2) the criteria of investment. To list up the priority of investment requires analyzing plant accident data and process data with the general safety information by using matrix computation method that we propose. To set up the criteria of investment requires surveying business-level executives regarding acceptable safety damage such as casualty, cost loss and business interruption.

Figure 5. Safety related Business Decision-Making Procedure.

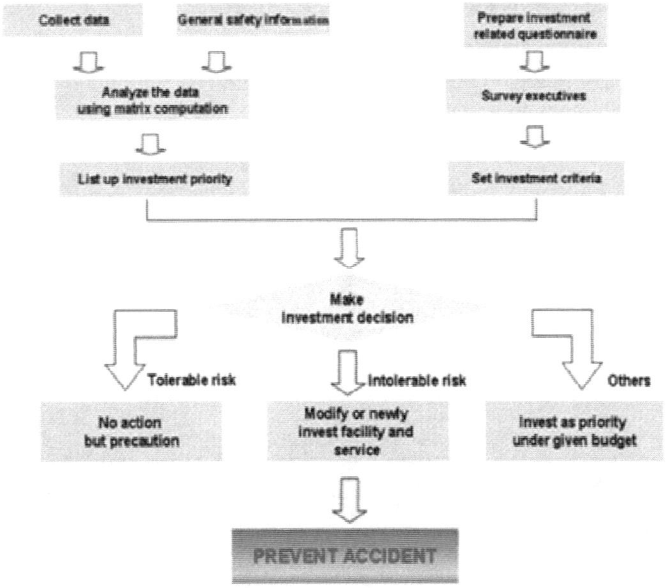

Based on the above two preparations such as the priority and the criteria, the decision is made according to the three risk probability regions: 1) definitely tolerable risk region, 2) definitely intolerable risk region and 3) other region in between the two. If the risk of a hazard is tolerable, the company does not need to invest any money into the relevant event and, in many cases, just careful attention is necessary. If the risk is definitely intolerable, the company must invest money to modify the process or purchase new facility and related service regardless of the budget if the plant needs to be run continuously. If the risk is between definitely tolerable and intolerable region, the investment is negotiable with the given budget that is, the investment to each event is selected by the priority under the condition that the given budget allows. As following this decision procedure, the company invests money only to necessary facility and service for preventing safety accident under the given budget.

7. Study Case: Surface Upstream Project

7.1 Scope

The whole Station Upgrade Project comprises the installation of new facilities, and revamp of existing ones needed to increase the produced water treatment capacity during production of Extra Heavy Oil (EHO) of 100 KSBPD to 200 KSBPD.

In more detail (Figure 6), MSUP includes:

- A new Diluted Crude Oil (DCO) Treating Train (Train C), capable of treating 95 KSBPD of EHO diluted with 34.4 KSBPD of Diluent, with 77 KSBPD of associated produced water.
- A new Produced Water Treating Plant of 200 KSBPD.
- Utilities revamp, including a new 25 MW Turbogenerator and 100 KBBL Diluent tank.
- Minor upgrades in existing DCO Treating Trains.
- New Flare
- New Fuel Gas Skid
- New Chemical Injection Packages
- New Desanding Systems
- New Open and Closed Drain System
- New Control Room (MSCR)

7.2 Location

Main Station facilities are located to 7,5 km South-Eastern from San Diego de Cabrutica, near from region Pariaguán, to South Anzoátegui State, in Venezuela.

7.3 Facilities Description

Main Station facilities are part of the Upstream Section of an overall heavy oil production and upgrading concept that will lead to the production of Synthetic Crude. Development of this concept is under way through two different Engineering Projects. The project that deals with the production and initial conditioning of the heavy oil is referred to as the Upstream Surface

Facilities Project, while the one that deals with the transformation or upgrading of the heavy oil to Synthetic Crude is referred to as the Downstream Facilities Project.

The major elements of the Upstream Section are:

- Oil Wells, grouped in Clusters, to produce the heavy crude.
- Diluent System, to dilute the heavy crude at the well head and improve its pumpability.
- Gathering Network, to collect and send the diluted heavy crude for initial conditioning.
- Main Station, to carry out the initial conditioning.

The facilities in the Main Station Upgrade Project are divided into Third Oil Dehydration Train called "Train C", a Production Water Treatment Plant (PWTP), and some utilities revamp (fuel gas, Diluent, open and closed drains).

The existing MS Oil Processing Trains (A and B) are not capable to handle the new production target 246 KSBPD EHO, 89 KSBPD Diluent and 200 KSBPD Water due to the increasing water cut. In order to handle the fluids coming from clusters a Third train is required.

- Export Pipeline, to send the conditioned heavy crude to an intermediate Pump Station, operated by another party that will finally get the oil to the Upgrader Section.

Main Station's primary objective is to process the incoming fluids from the gathering network, i.e. oil, associated gas and water, so that oil can be consistently exported to the Upgrader within the required specs and at proper pumping conditions.

Oil to be exported from Main Station shall have a maximum water cut of 2%, a Reid Vapor Pressure no greater than 0.76 bar_a (11 psi_a), a specific gravity of 0.953 (17° API) and shall be pumped at approximately 89 °C and 25 barg.

The crude oil is separated from the gas and is stabilized to less than 0.76 bar_a (11 psi_a) RVP and the water content is reduced to less than 2% vol. The API gravity of the crude oil is adjusted by adding diluents to 17 °API and is then pumped to the export pipeline. The facilities at the Main Station are continuously manned.

The fluids coming from the gathering network enter Main Station and split into two identical trains at the upstream of the Slug Catchers (SC) where most of the associated gas is separated and sent to the fuel gas system.

The liquid from the SC is heated through heat exchangers and direct fired heaters so it gets to the right temperature to release enough light ends (gas) to meet the RVP spec in the High Temperature Separators (HTS). The vapor released in the HTS is cooled down so the diluents fraction that evaporated in the HTS is condensed, separated from water and returned to the export pump suction. The remaining gas is sent to the fuel gas system and the water is sent to treatment.

The new PWTP is located at Main Station and it is designed to process 1325 Sm^3/h (200 KSBPD) of production water. Treated water has oil content below 30 ppm and total solids content below 20 ppm making it suitable for re-injection.

The PWTP consists of the following water clarifying equipment:

- Skimmer Tanks T-3900 C/D and associated Gas Boots D-3900 C/D (working in parallel), with associated pumps: Skimmer Tank Transfer Pumps, P-3901 A/B/S (P-3901 A for T-3900 C and P-3901 B for T-3900 D; P-3901 S is a common spare) and Oil Recovery Pumps, P-3951 A/B/S (P-3951 A for T-3900 C and P-3951 B for T-3900 D; P-3951 S is a common spare).
- Flotation Units X-3900 B/C with associated Flotation Transfer Pumps, P-3921 A/B/S (P-3921 A for X-3900 B and P-3921 B for X-3900 C; P-3921 S is a common spare).

- A Flotation Overflow Separation Tank T-3930 B, with associated pumps: Flotation Recycling pumps P-3926 A/S and Flotation Sludge Recovery Pumps P-3961 A/S.
- A Clean Water Tank T-3920 B to store filtered water and,
- High - pressure Water Re-injection Pumps P-3932 A/B/S.

Utilities such as Fuel gas, Firewater, Instrument Air, Plant Air, Drinking and Utility Water, Diluents required by Train C and PWTP are provided by Main Station.
The Fuel Gas used in Main Station it is provide by the incoming fluids from clusters through Trains A/B/C (Gas Oil Ratio between 50 - 250 SCF/BBL, about 12,3 MMSCFD – 61,5 MMSCFD), and the one coming from the Gas Gathering Network, for a maximum gas flow rate of 120 MMSCFD, according to gas production forecast by MSUP Study.
In order to handle this peak flow rate new Fuel Gas facilities and LP Flare System will be enhanced.
All this gas is sent to the Fuel Gas KO Drums, D-3201 A/B/C, and Fuel Gas Filters F-3201 A/B/S, and then it is distributed to the users, namely:

- Oil Heaters, H-3001 A/B/C,
- Power Generation Units, GT-3201 A/B/C/D,
- Off-spec Storage Tank, T-3002 and the Production Water Treatment Plants for blanketing purposes.
- Flare system (LP and LLP), for continuous purge.

Gas for turbogenerators is compressed in the Fuel Gas Compressor Packages, K-3201 A/B/C/S while the heaters and relief system can use it at the pressure level supplied by the KO drums.
Firewater, Instrument Air and Utility Air, Utility Water and Drinking Water connection points to Main Station are listed in Tie-in Point List. Electrical power is supplied by Turbo Generators, GT-3201 A/B/C, for MSUP a new Power Generation Unit, GT-3102D, will be installed, with its associated equipment (new generator substation G/S-02 plus an electrical substation T/S-04 dedicated to New Oil and Water Processing Trains). Additionally MS has two Emergency Diesel Generators, EDGS-1/2.
The PWTP normal operation is manual but it can be monitored from the Main Station Control Room and by sample testing at different points throughout the Plant.

7.4 Risk Analysis - HAZOP

The key to the Process Hazard Analysis (PHA) is a simple concept offers ways to answer the following four (04) questions, also a greater risk reduction, beyond the goals above, may be appropriate where an analysis cost-effectiveness of permits:

a. What can go wrong?
b. What are the causes?
c. What are the consequences?
d. What is the probability of occurrence?

The implementation of the entire set of techniques PHA comprises a quantitative review risks (Train C, New Water Treatment Production Water (NWTPW), Generation System and Common Services including here: Fuel Gas System, Low Low Pressure and Low Pressure Relief System, Diluent System, Open and Closed Drain Systems, Instrument and Plant Air System, Potable and Service Water System, Chemical Injection Systems) that may be in an installation, ranging from frequent incidents low-stakes to major incidents rare and, using a methodology uniform and consistent.

Once identified risk processes, the techniques can PHA help focus control studies risk. The PHA must address the following:

- The hazards of the process
- Identification of any previous incidents with had a likely potential for catastrophic consequences
- Engineering and administrative controls applicable to hazard, their interrelationships and consequences of failure of these controls, especially those affecting employees and environment
- Facility siting
- Human factors
- Evaluation of effect of failure of controls
- The requirement to promptly resolve PHA findings and recommendations

The largest contributors to risk can be identified and can be recommendations and take decisions, as corrective measures, on the basis objective and consistent.

The business value of the hazards analysis is the use of report to train people on how deviations from normal operation propagate to process hazards. The risk analysis defines the risk reduction strategy selected to address unacceptable risk. Plant management, engineering, operations, and maintenance must understand this information if they have responsibility for decisions affecting the hazards or risk.

Operator and maintenance procedures should include a description of the process hazards where their actions potentially increase risk. The hazards analysis documents should be discussed at Operations' safety meetings.

Residual risk should be addressed with compensating measures and action plans to reduce risk as required.

In this case, don't was identified the hazard about:

- Wrong specification for suction and discharge valves at Fired-direct Furnace, H-3001C, (Figure 8)
- Lack of control for valves at inlet pipelines
- Among others

Figure 8. Fired direct Furnace H–3001C

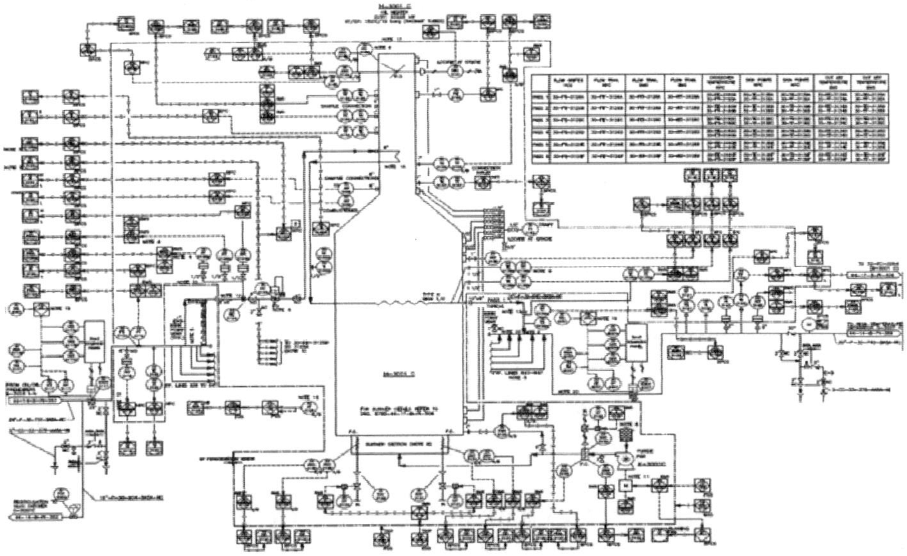

7.5 Allocation of Requirements of Safety Integrity Level (SIL) of Safety Instrumented System (SIS)

Due to the worldwide major accidents, most industrialized nations established process hazards control regulations. Germany passed the Hazardous Incident Ordinance (1980) and the European Community created the Major Accident Hazards Directive (1982). The United States passed legislation creating the Emergency Planning and Community Right-To-Know Act (EPCRA, 1986). In 1990, the Clean Air Act Amendments mandated that the Occupational Safety and Health Administration (OSHA) and the Environmental Protection Agency (EPA) develop accident prevention regulations, leading eventually to the 1992 OSHA Process Safety Management (PSM) and the 1996 EPA Risk Management Program (RMP) regulations.

None of the regulations were prescriptive. Most were intended to be flexible, simply requiring industry to examine the risk posed by their processes and to work to minimize these risks. Only a few European regulations provided quantitative risk targets. Many European regulations, as well as those of the U.S., relied on industry to develop its own tools to identify risk and to make appropriate efforts to achieve lower risk. As an incentive, industry was required to communicate the identified process risks to the public and emergency responders through public forums and documentation.

The industrial community responded to the regulations by issuing industry standards and guidelines concerning the evaluation of risk and the design of safety systems. The American Petroleum Institute (API), the American Society of Mechanical Engineers (ASME), and the National Fire Protection Agency (NFPA) created codes and practices for specific applications. In 1988, the International Society for Measurement and Control (ISA) began an eight year odyssey to develop a standard for safety instrumented system design (SIS) for the process industries, ANSI/ISA 84.01-1996; late In the mid-1980s, the International Electrotechnical Commission (IEC) began the development of an international standard for the design of all safety-related systems, covering transportation, medical, manufacturing, and process industries as IEC 61508. And then at 2003, ANSI/ISA.S.84.00.01-2004 and IEC 61511.

Consequently, it is regulations the mandate that industry identify and address risk. Industry standards are simply providing the tools for assessing the adequacy of industry's efforts. The difficult task of creating one standard for all of the process industry made the creation of a prescriptive standard impossible. After all, could one prescriptive standard cover refinery furnaces, chemical reactors, pulp and paper digesters, and utility boilers and reduce risk appropriately at a reasonable cost? Thus, the standards had to be performance based. Both standards chose to rely on the establishment of a design process called the SIS lifecycle, throughout which the performance of the instrumented systems must be maintained.

The lifecycle is intended to address the primary causes of control system incidents. Compliance with the industrial standards requires four essential elements:

a. Identification of safety functions required for safe shutdown;
b. Assignment of a safety integrity level (SIL) for each safety function;
c. Use of the safety lifecycle for the SIS design; and
d. Verification of the SIL achieved for each safety function.

These elements result in a major paradigm shift for SIS design. While it has been accepted practice for many years to mitigate potential incidents with instrumented systems, there has generally been no assessment of what type of SIS provides the appropriate risk reduction. These standards have now established requirements that industry document the rationale behind the use of a particular design, evaluate that design for its integrity, and demonstrate that the integrity can be maintained. This is perhaps the most significant impact of ANSI/ISA 84.01 and IEC 61508 on process safety, in general.

As performance based standards, the race for compliance is marked with the broadly defined finish line of good engineering practice and highly recommended practices. As industry races toward compliance, it must work hard to prevent the creation and acceptance of bad engineering practices, which threaten the economics of plant operation and erode the effectiveness of SIS designs, addressing seven bad engineering practices:

1. Believing that, if something is not specifically stated, either "shall do" or "shall not do," in the standards; you do not have to worry about it.
2. Thinking that meeting the minimum requirements means the process is safe and the SIS is compliant with the standard.
3. Ignoring the importance of good engineering practice.
4. Designing systems that meet safety requirements but not economic protection requirements.
5. Focusing only on SIL and not on preventing nuisance trips.
6. Neglecting the human factors.
7. Focusing on capital cost and not lifecycle costs.

A special attention should be paid to the possible common mode causes which may induce a failure of several (or all) barriers of the protection system. They include namely:

- The power supplies of the system.
- The logic control system(s)
- The valves activation system(s)
- The corrosion/erosion/plugging effects of the crude fluid.
- The operation of the system.

The Active Protection System is to be based on different independent barriers acting at different levels in case of an escalation of any deviation from the normal operation:

- Alarm: provides a warning to the operators in order to take a corrective action, given the deviation is not expected to become hazardous during the performance of the action.
- Production Shutdown System (PSDS): avoids the occurrence of an accidental event upon detection of a hazardous situation. The related action induces generally the shutdown of the related equipment through the implementation of one or several of the following barriers:
  - Primary barrier: aims at suppressing the cause of the deviation.
  - Secondary barrier: aims at limiting the effect of the deviation onto the equipment.

The PSDS may include several sub-levels, given there is General production Shutdown level initiating the shutdown of the whole production equipment.

Emergency Shutdown System (ESDS) triggers the safety shutdown of the installation upon detection of, either an **accidental event** or of a hazardous situation which may escalate rapidly. The ultimate ESD level causes the isolation of the installation at its limits through ESDV's and SCSSSV's (wells). It initiates the General production Shutdown and the shutdown of the related utilities. It keeps the necessary resources for emergency response. The ESDS may include several sub-levels, given the ultimate level, as defined here above, has priority on the other levels.

The Industry Practice regarding the control of the overpressure risk is to consider, by order of priority, the following possible options:

1. Design the equipment and/or the pipeline system to sustain the maximum possible pressure which may occur in the equipment (safety by design).
2. Design a protection system which, in accordance with API RP 14C, is made of:
    - A primary barrier aiming at isolating the heated equipment from the source of overpressure (suppressing the cause).
    - A secondary barrier aiming at relieving the pressure through a suitably sized venting or flaring system (limiting the effects).
3. Design an instrumented protection system made of two independent barriers acting at different levels to isolate the heated equipment from the source of overpressure before the pressure exceeds the maximum allowable pressure. This option named IPPS is considered when it is shown the options (1) and (2) induce disproportionate costs, technical or environmental problems. When a rapid escalation of the overpressure is expected, the IPPS may trigger the closure of the SCSSV's and/or ESDV's through the ESDS, in that case, the ESDS provides an additional and independent barrier.

The system should be designed fail-safe which means that the system will revert to a predetermined safe state in the event of failure of its components or of its power supplies. These include:

- Loss of the actuation energy of the valves.
- Loss of the control signal of the valves.
- Loss of the overpressure detection sub-system.
- Loss of the logic sub-system.

The overpressure protection system is to be based on the concept of two independent active barriers automatically initiated, at different levels, by the early detection of the pressure built-up into the protected equipment.

The system should survive any single failure of its components without jeopardizing its protection function. This is to be implemented through the research and elimination of the possible common mode failures, implementation of redundancies or by fail-safe action. Consideration should be given in the design of the redundancies to the using of different technologies (electrical/hydraulic valve activation as example).

The system has to be designed in order to reduce the probability of unrevealed failures of its components. This is achieved by implementation, where applicable, of monitoring and/or automatic testing of the components triggering an alarm in case of detected failure.

The system has to be designed in order to facilitate the periodic testing of the system. Consideration should be given to the possibility of performing a partial closure and re-opening of the valves while producing when this improve the availability of the protection system.

The overpressure protection system including the dedicated IPPS components should be considered as a whole system regarding its dedicated target and general objective. The reason of this approach is to facilitate the control of the overpressure hazard by considering all possible causes and their effects and namely to tackle the possible common mode failures which might involve a failure of the whole system, for example, for fired-direct furnace H-3001C it is considered an overpressure protection system that don't operate because the shutdown valves (input and output; continuing service) are specified as "Fail Last Position" and then, occurring an event will not protect the equipment, operators, assets and environment.

7.5.1 Evaluated Systems as New Train C, New Water Treatment Production Water (NWTPW), Generation System and Common Services including here: Fuel Gas System, Low Low Pressure and Low Pressure Relief System, Diluent System, Open and Closed Drain Systems, Instrument and Plant Air System, Potable and Service Water System, Chemical Injection Systems

In general, SIL3 is configured on all systems which is not sufficient safety to conduct a risk that has negative consequences and high exposure to high toxic concentrations.

Provided the single safety function (SIF) as HAZOP results, you're talking about lies within the scope of IEC 61511/62061 (i.e., a SRECS for instruments and machinery), and – as long as it functions properly – is sufficient to reduce the residual risk to an acceptable level, a reliability level of SIL3 for that single SIF is sufficient. By design that single SF will have to include sufficient redundancy/diversity/monitoring to be able to achieve SIL3.

With I mean that a if the single SF concerns e.g. the guarding of a shutdown valve to a danger zone, no other means of access to that same danger zone should exist, unless these are also guarded by SIL3 safety functions.

Remember that according to the principle of integrated safety for equipments, a SIS should only be considered "after" the possibilities of design changes and of physical protection measures have been exhausted. If the residual risk at the end is still too high, should consider additional constraints on the intended use of the equipments.

If use a simple pressure switch (diaphragm to lever-arm to electrical switch) in a safety interlock circuit where PL_a is called for, is a standard off-the-shelf pressure switch appropriate?, and If PL_c is called for (due to higher pressure), is still; possible to use a high reliability pressure switch, or would one use a pressure transmitter in a SIL1 safety function.

Can use a commercially off the shelf pressure switch provided it is a well tried component with known failure modes for a PL_a safety related control function, however to achieve PL_c, you will need to use redundant pressure switches. Using a single pressure switch may not give you the desired reliabilities due to HFT = 0 using IEC62061 or even with the use of ISO13849-1 due to a single channel circuit. However, it is important to bear in mind that the safety related control function extends from the sensing switch right up to the actuating device; and the lowest PL or SIL in the chain will influence the final result for the entire safety loop. A pressure vessel would fall under the pressure equipment directive 97/23/EC and the IEC 61511 functional safety standard for such systems would take over. Principles are similar using SIL too.

PL_c capability can be achieved in a single channel subsystem providing that the components have high $MTTF_d$. For a SIL1 according to Table 5 in IEC 62061 it is possible to achieve SIL1 with HFT = 0 providing that SFF is >60%. Single channel sensor subsystems can also achieve SIL1 according to Table 6 in IEC 61511.

The common cause and systematic failures that affect even a SIL Logic Solver will affect all protection layers that the logic solver is part of, and these will as a consequence of this not be independent. In fact this is subject to discussion, where some are of the opinion that a SIL3 logic solver is able to cover both the 2 IPLs required by ISO 10418.

These SIFs would not qualify to be IPLs. However, I would think that the SIFs should reinforce the protection towards the same risk, especially if the Logic Solver has very low failure rate compared to the field devices.

The independent rule would be hurt. According Layer of Protection Analysis, CCPS 2001: "In order to be considered an IPL, a device, system, or action must be independent of the initiating event and the components of any other IPL already claimed for the same scenario…" However, one of them could be considered as a safeguard, none SIL claimed.

Depends on having independent paths through the SIL3 Logic Solver. It is possible to have two separate SIF's with different initiators and different final elements to cater for the same risk but same logic solver (SIL3 rated) are these IPL's.

Adding SIFs for the same risk don't provide you with more IPL because:

- They aren't independent, because even they use different instruments; they share the same logic solver.
- They aren't different IPL, because both of them are SIF. An extra PL could be a mechanical relief valve, and plus, it will be independent from the logic solver, so you will have another IPL. That way if the inner previous IPL doesn't work (you can consider your SIF as this inner IPL) you will have another extra IPL.
- The common cause and systematic failures of the SIL3 Logic Solver will affect both protection layers. Therefore the LS IPL load is limited to 3, one could load a 2 SIL1 loops or 1 SIL1 + 1 SIl2 on this system, but one could NOT load 2 SIL2 loops on this system. This according IEC 61511-part 1, clause 9.5.1, part 2 Clauses 9.5.3.
- As you know that SIS consists of SIF's and you can use single or redundant logic solver for those SIF's (in SIS). So my answer for using same logic solver for many SIF is yes you can, single/redundant LS will be depend to your verification that you will count many things (diagnostic covered, beta factor for common cause, etc) to make sure that your SIF equipment reliability will meet level target (SIL target / PFD).

A Common Cause Failure (CCF) occurs when a single fault results in the corresponding failure of multiple components, such as a miscalibration error on a bank of redundant transmitters. The frequency of common cause faults is difficult to estimate. The modeling techniques and available failure rate data make the predictive calculations of these failures cumbersome and the results obtained questionable. Therefore, a more meaningful way for most SIS designers is to eliminate the potential source of CCF in the SIS design, installation, operation and maintenance.

The SIS should be designed to meet the required safety integrity level as defined in the safety allocation requirement specification. Moreover, the SIS design should be performed in a way that minimizes the potential for common mode or common cause failures. A CCF occurs when a single fault results in the corresponding failure of multiple components. Thus, CCF's can result in the SIS falling to function when there is a process demand. Consequently, CCF's must be identified during the design process and the potential impact on the SIS functionality must be understood.

Unfortunately, there is a great deal of disagreement among the experts on how to define CCF and what specific events comprise a CCF. The following are often cited as examples of common cause faults:

- Miscalibration or no calibration of sensors
- Pluggage of common process taps for redundant sensors
- Incorrect maintenance or no maintenance
- Improper bypassing
- Environmental stress on the field device
- Process fluid or contaminant plugs valve

But the examination of these faults, in light of any SIS design, will indicate that any of these six examples can disable single I/O systems, as well as redundant I/O systems. However, many of the proposed methodologies for assessing CCF ignore this fact and only penalize redundant sensors and final elements.

The ultimate failure of all is that the safety requirement specification (SRS) is incorrect at the beginning of the design process and the transmitter cannot detect the potential incident. This is the most disastrous common cause failure that directly leads to the hazardous incident that the designer is seeking to prevent. A bad SRS is the ultimate "trump card" for the entire SIS and is one failure that most of the proposed methodologies ignore.

The choice of the evaluation technique is typically dependent on experience of the User with the particular SIS design. This would include documented historical performance of instruments, installation details, and design engineering teams. Experience with the specific application environment is required, because a device or installation detail that works well in one type of application may not work well in another. For example, standard taps into a vessel for mounting a transmitter may work extremely well in clean service, but may plug very quickly in a service where solids can deposit. Three qualitative techniques are often used to assess SIS designs:

- Industrial Standards
- Engineering Guidelines and Standards
- Qualitative Assessment as Checklist

8. Conclusions and Recommendations

These new rules require a change of focus for everyone associated with these systems. The change of focus will result in a significant learning curve for many engineers. In order to make this transition as smooth as possible, rigorous qualitative techniques should be employed to provide a method for assessing these systems for adequacy. Clear communication of the requirements for effective SIS design is essential. All procedures, internal guidelines, and checklists must be evaluated by all personnel involved to ensure a compliant and cost effective SIS.

One suggestion that is obvious but difficult due to human nature is – help management (refinery/corporate) to truly see the OSHA inspection as a positive learning opportunity.

An efficient, accurate OSHA inspection can provide direction to upgrading a refineries PSM program into one that supports sustainability. Long term maintenance of a rigorous refinery PSM system will prove its business case in terms of:

- Corporate Responsibility
- Business Flexibility
- Risk Reduction
- Sustained Value

An effective PSM system upgrade resulting from the inspection can actually increase a refinery's reliability and ability to sustain long term success.

The refining industry will have a directed opportunity to improve overall performance in process safety due to the OSHA refinery process safety management national emphasis program. Leveraging this opportunity requires a culture that views personnel and process safety as second nature – a conditioned first response to address for any issue related to refinery operations or maintenance.

High reliability organizations (such as refineries, petrochemical complexes, chemical plants and power plants) whose upper management values long term growth and more predictable physical plant reliability will embrace the chance to use the effort involved in a PSM program implementation based upgrade to bolster these benefits.

We now have benefits and costs. How are these converted into a proper overall project financial analysis? How can we demonstrate that the proposed investment will increase the long term corporate ROIC more than other potential investments?

The cultural level of operational discipline at the corporation, team, and individual level can better support a high level of operational excellence – doing those critical measurable things that support success – when a program is detailed and monitored. Business excellence can be attained when a company values a holistic sense of what comprises quality.

The Figure 9 represents this view. OSHA inspectors and management have gained a new level of process safety management expertise based upon their roles in the Industry. This can only enhance OSHA's ability to help industry protect the workforce from preventable risk. If production facility management adopts this view, the industry's level of catastrophic risk prevention will increase.

Figure 9. The Holistic View of Quality

Although the specific investment evaluation protocols vary from company to company and you need to follow your company's standards, they all have at their center the calculation of the net present value of the incremental after tax benefits generated by the plant due to the equipment or technology and comparing it with the net present value of the required investment. In financial terms, all assets, be they stocks, bonds or plants, are valued the same, i.e. as the net present value of the future sequence of after tax cash flows. If the value is greater than the investment the project is considered for funding. Net present value implies discounting a sequence of net cash flows (receipts less expenditures) back to the current time. But there are many subtleties.

The Operation and Maintenance staff of today must justify how the SIS is operated, when it is bypassed, when it is tested, and when maintenance is performed.

9. Acknowledgement

The author would like to thank our Meeting assistant colleagues for many discussions, which assisted in the evolution of the ideas presented herein to:

AIChE's 2011 Annual Meeting, United States of America, October 16-21, 2011, Minneapolis, MN, and

inelectra Operations VicePresidency (Eng. Milagros Sulbarán) and Engineering Management (Eng. Victoria Granados), for their cooperation and support.

11. Author: Glenda J. Gutiérrez Capriles

She was born in Valencia, Carabobo, Venezuela on January, 1974. She obtained her B.Sc. from Los Andes University (ULA) in Merida - Venezuela, in 1996. Then in 1999-2000 obtained her title as Magisterium Scientarium Specialist in Environmental, Health and Safety, ULA.

Also she holds certification in PMP and TUV FS.

Between 2000 – 2008 she served from Process Specialist, Instrumentation Engineer, Risk Engineer, Maintenance Centered in Reliability Supervisor, Operational Engineering Superintendent for PDVSA E&P, PDVSA GAS and PEQUIVEN.

Gutiérrez Capriles is currently worked as Project Risk Engineer for inelectra Venezuela from beginning of 2009.

Gutiérrez Capriles has over fourteen (14) years of experience as Engineer, Consultant, Operator, highlighting those regarding Risk Studies as Inherently Safer Designs, HAZID, HAZOP, SIL/SIS, FTA, ETA, FMEA, Reliability Analysis as RCM, RBI, OEE, RAM, RAMOS or MORAS, Project Risk Analysis as LCC, DSD, PEP, among others.

She is currently worked as Project Engineer for inelectra Venezuela from the beginning of 2009. She has over fifteen (15) years of experience as engineer, consulter, operator, highlighting those regarding risk studies, reliability analysis, among others.

Her experiences in Process Safety Knowledge are cited on many national and International Journal and Seminars. She has published and presented papers at several events organized by relevant institutions at national and international levels.

- 2010 Chemical Engineers Argentine Congress, Mar del Plata, Argentina.
- 3rd CCPS Latin America Conference and Expo, Buenos Aires, Argentina.
- 2011 Annual Meeting AIChE, Minnesota, USA.
- AIChE Journal, March 2001 Performance Assessment of Constrained Model Predictive Control Systems.
- Gutiérrez-Capriles G. "I Jornadas de Protección Integral". Universidad de Los Andes. Facultad de Ingeniería. Mayo 1996.
- Gutiérrez-Capriles G. Ergonomía en los diferentes Ambientes de Trabajo (Tesis de Magister). Mención Especial de Honor por la Sociedad Venezolana de Medicina Ocupacional, Escuela de Salud Ocupacional de la Facultad de Medicina y Escuela de Protección Integral de la Universidad de Los Andes. Universidad de Los Andes. Mérida, Venezuela (1999)
- Gutiérrez-Capriles G. Anteproyecto del Diseño de una Planta para la Obtención de Lecitina Blanqueada de Soya (Diseño de Plantas). Memorias del II Congreso Nacional de Estudiantes de Ingeniería Química. Universidad de Los Andes. Mérida, Venezuela (1999)
- Gutiérrez-Capriles G. Peña A. BATCH: Programa de Computación simulando una Torre de Destilación por Lotes Batch. Memorias de las IV Jornadas Científico-Técnicas de la Facultad de Ingeniería. Universidad de Los Andes. Mérida, Venezuela (1998)
- Gutiérrez-Capriles G. "III Congreso Internacional de Educación Superior en materia de Drogas". Universidad de Los Andes. Abril 1996.
- Gutiérrez-Capriles G. "I Congreso Internacional de Educación Integral en materia de Seguridad sobre Riesgos Sísmicos". Vicerrectorado Académico y Brigada de Rescate de la Universidad de Los Andes. Abril 1996.
- Gutiérrez-Capriles G. Optimización de Mezclas de Policloruro de Vinilo utilizadas para la Elaboración de Envases Plásticos (Tesis de Grado). Mención Especial de Honor por la Sociedad Venezolana de Ingenieros de Polímeros, Departamento de Química de la Facultad de Ciencias y Departamento de Ingeniería Química de la Universidad de Los Andes. Universidad de Los Andes. Mérida, Venezuela (1996)
- Gutiérrez-Capriles G, Belandria J. Graficador Termodinámico de Compuestos Químicos Puros en Fase Fluida por 4 diferentes Métodos Termodinámicos. II Congreso Interamericano de Computación Aplicada a la Industria de Procesos. Santiago de Chile, Chile (1994)

(512c) Exogenous Uncertainty: Choosing Innovation Projects

Diana Matonis, Flucomp Enterprises, Hickory Hills, IL

What is the difference between risk and uncertainty?

What are some quantitative methods that aid decision making under uncertainty?

What are some qualitative methods that aid decision making under uncertainty?

How can decision makers effectively combine qualitative and quantitative methods?

See more of this Session: Challenges In Open Innovation and Growth
See more of this Group/Topical: Management Division

Paper Unavailable

(512d) Identifying & Transforming Local Best Practices to a Global Competence

Marcus A. Tennant, Yokogawa Corp. of America, Pleasant Prairie, WI

Often in global corporations, best practices emerge that enable a company to improve business performance at alocal facility or region. In this presentation we will look at how best practices emerge locally but are often ignoredin a global perspective due to their local or regional origin. Many of these best practices are developed "below theradar" and work very well for a local market; however small changes and the proper strategy for adaptation canturn the best practices into company-wide global competence. We will we will discuss methods that a companycan "dig out" these practices and find the critical success factors, discuss strategies of implementing these bestpractices in other regions and how they could be used to improve business performance globally. Finally, we willalso look at potential pitfalls and how to overcome obstacles in a global implementation effort.

See more of this Session: Challenges In Open Innovation and Growth
See more of this Group/Topical: Management Division

AUTHOR INDEX